Quantum Ecology

Quantum Ecology

Why and How New Information Technologies Will Reshape Societies

Stefano Calzati and Derrick de Kerckhove

The MIT Press
Cambridge, Massachusetts
London, England

The MIT Press would like to thank the anonymous peer reviewers who provided comments on drafts of this book. The generous work of academic experts is essential for establishing the authority and quality of our publications. We acknowledge with gratitude the contributions of these otherwise uncredited readers.

This book was set in Stone Serif and Stone Sans by Westchester Publishing Services. Printed and bound in the United States of America.

Library of Congress Cataloging-in-Publication Data is available.

ISBN: 978-0-262-54621-8

10 9 8 7 6 5 4 3 2 1

Contents

Introduction

Two cultures or technologies can, like astronomical galaxies, pass through one another without collision; but not without change of configuration.
—Marshall McLuhan (1962, 149)

Why a Quantum Ecology?

Claims that consider the present as a point of rupture with the past have been made throughout history, especially in connection with radical societal changes—from wars and pandemics to economic crises and political revolts across the globe—and/or technological innovation—just think of the consolidation of mass media during the twentieth century down to the commercial outbreak of the internet in the 1990s and now the emergence of increasingly refined artificial intelligences (AI). As soon as one takes a long-term perspective over these phenomena—however disruptive they might appear at first sight—their innovative character is often smoothed out and somewhat swallowed by history. The development and spread of quantum technologies[1] might be no exception in this regard. And yet, there is a galaxy of converging social, political, cultural, and technological signs that also indicate how humanity might be actually transiting into a radically different epoch. How should the overt and hidden social, cultural, and political tensions, at both global and local levels, be interpreted? What is the nature of these tensions? How to tackle them? Of course, in perspective, one could see these tensions too as ordinary historical bumps, but their scale and depth suggest otherwise.

The main argument of this book is that humanity is entering a new sociotechnical paradigm, with quantum technologies as a pivotal *dispositif*

(Foucault 1980) in defining and shaping such a paradigm. The main difference with the past is that these technologies rely upon, embed, realize, and disseminate at large scale quantum mechanics'[2] counterintuitive behaviors and principles—such as, superposition, entanglement, uncertainty, discreteness—which question the existence of a reality independent from the knower. Starting from the Heideggerian (Heidegger 1977) idea of technology as a certain *enframing* of experience, we claim that emerging quantum technologies—especially quantum information technologies (QITs)—will redesign human experience based on these behaviors and principles, reworking the classic understanding of the world people have and how they act in it, according to language-based technologies, from the press to mass media and the internet. The shift from a language-based to a quantum-based enframing—or what we call *ecology*—is neither definitive nor clear cut; rather, it manifests a reshuffling of values: epistemological, ethical, and aesthetic, as well as political. In this shifting process, we identify today's digital transformation[3] as the middle-step ecology—based on massive data production and algorithmic processing—bridging the other two. At stake, then, is not the disposal of one ecology over another but rather the exploration of the (un)balanced power relations among the three and the effects these produce on each other. More broadly, we outline the contour of what we call a *theory of open reality*: quantum physics tells us that a unique, objective reality does not exists; what exists—we claim—are *dispositif*-dependent ecologies, intended as *sociotechnical processes* that shape *self-organizing world-sensing*—enacted through *embodiment*, operationalized via specific *operating systems*, and articulated in different *technocultural fields*—which repeatedly overlap and conflict within and across themselves, coupling *individual* and *collective* dimensions in a mutually implicated order of emergence. Such a claim certainly begs for explanation.

*   *   *

As it might be already clear, this book is not about quantum mechanics per se—we leave this to professional physicists; rather, from/through quantum physics, we outline the emergence of a new sociotechnical paradigm gravitating around quantum technologies and enacting quantum behaviors, principles, and phenomena at the systemic level. To be sure, we do not posit any deterministic link between quantum technologies and societies: indeed, as we will explain, we do keep the cultural specificity of different contexts very much into account, and we consider technology not so much in essentialist terms but always as a *dispositif*, that is, the aggregation of tangible and intangible assets and values. Nor do we want to posit loose metaphors between quantum physics and what is happening today in the world: our reading of quantum physics, while being based on scientific evidence, points chiefly to the philosophy at its basis. What we do claim, instead, is that quantum technologies will generate a paradigmatic shift—extending the line that connects language and digitalization—which demands new attitudes, interpretative frameworks, and courses of action to be coped with. Our goal throughout the book will be to unravel the roots of such a shift and provide (nonprescriptive) guidance on how to navigate the emerging quantum ecology.

At a fundamental level, then, this book unpacks the human–technology relation exploring what it means to communicate, to be human, to know, and, ultimately, to act in the world—to be living. Equipped with an understanding of the debates surrounding the interpretation(s) of quantum mechanics, we will investigate these deep onto-epistemological issues starting from the tradition of media studies—especially the informal Toronto School of Communication, of which Derrick has been a major voice for over thirty years—and transitioning toward sociology, digital cultures, and the (geopolitical) governance of AI and digital transformation, which are subjects at the core of both authors' research and teaching in different contexts and times—but especially, together, at the Polytechnic Institute of Milan. Due to the breadth of the issues we will address—which is inevitable given the fundamental nature and impact of quantum physics—the book will reach out toward other fields, benefiting from insights from areas as diverse as philosophy and anthropology, critical data studies and evolutionary biology, as well as information theory and cosmology. While abiding to a rigorous treatment of the argument, we strongly believe in the need to pursue not only interdisciplinarity but also transdisciplinary thinking and research. Paraphrasing the great sociologist Max Weber, we should have no prescriptive fields; we are not donkeys. This, too, is a major teaching coming from the quantum revolution, as computer scientist Scott Aaronson (2013, xvii) explains well: "While quantum mechanics was invented a century ago to solve technical problems in physics, today it can be fruitfully explained from an extremely different perspective: as part of the history of ideas, in math, logic, computation, and philosophy, about the limits of the knowable."

At the Heart of Quantum Mechanics

Since its inception at the beginning of the twentieth century, quantum mechanics, which studies the subatomic world, has puzzled scientists by showing "oddities" and counterintuitive behaviors of matter's particles and light, when compared to the everyday macro-level reality. But what are these behaviors and principles? Here we introduce the most relevant ones for the present discussion, being aware that, while physicists today have gained considerable knowledge on *how* quantum physics works, there is still no full agreement on the *why*, that is, on the mathematical interpretation of several aspects of quantum mechanics.[4]

Planck's constant (and energy–matter discreteness) In his historical account retracing the birth of quantum mechanics at the beginning of twentieth century, Manjit Kumar (2008, 323) defines Planck's constant as follows: "A fundamental constant of nature that lies at the heart of quantum physics. Because Planck's constant is not zero, it is responsible for parsing, quantising, energy and other physical quantities in the atomic realm." In the nineteenth century, various experiments investigated the nature of light with the goal to settle the long-lasting debate about it being either a wave or a particle. Some of these experiments seemed to validate the wave hypothesis (against the corpuscular hypothesis held, for instance, by Isaac Newton).

At the exact turn of the twentieth century, however, Max Planck's studies on the electromagnetic radiations emitted by a black body opened a Pandora's box. A strenuous defender of the idea of continuous energy and matter, Planck worked toward the definition of a formula that could account for the observed discrepancy between theory and experiments as far as the intensity of radiation in the infrared region was concerned. In order to match data with theory, Planck was ultimately pushed to conclude that energy was emitted not uniformly but in packets or, more properly, in quantized form, the smallest unit of which has the value of what is known today as Planck's constant. This identifies the smallest value energy can take, setting the lowest threshold for any measurement, that is, the minimal amount of energy always involved in any observation.

Double-slit experiment In 1801, Thomas Young was the first to conceive the modern version of the double-slit experiment. He used sunlight—which he made pass through a hole split in two using a paper card—in order to study light's behavior. When the ray of light reached the screen behind the hole, it showed interference patterns: "When the two newly formed beams are received on a surface placed so as to intercept them, their light is divided by dark stripes into portions nearly equal" (Young 1807, 463). With this experiment, Young thought to have proved the wave nature of light.

However, on the wake of Planck's remarkable discovery, a number of similar experiments were conducted, ultimately showing that light manifests a dual nature, behaving sometimes as a wave and sometimes as a particle. The most known experiment is based on the following arrangement: a beam of light (photons) is directed to traverse two slits close to each other and hit a plate behind them. When firing one photon at a time, each photon that reaches the plate marks a particular spot on it. Light behaves like a particle. Yet, if one repeats the experiment over time, keeping track of photons' impact on the plate, interference patterns appear, consisting of a series of blurred bright and dark stripes. Light behaves also as a wave. A further version of the experiment includes the possibility of having a detector by the slits in order to track which slit each photon passes through; in this case, interference pattern is defused, questioning what makes photons behave like particles or a wave.

At the beginning of the twentieth century, studies on the atom, among which those by Niels Bohr, had already shown that the atom was not a monadic indivisible entity of matter; rather, it consisted of a nucleus—including protons and neutrons—and electrons gravitating around it. In 1924, physicist Louis de Broglie found that the wave-particle duality of light also applies to matter, by hypothesizing—his claim was later proven right—that subatomic matter's particles also behave like a wave. Put differently, particles such as electrons are observed to produce interferences, for example, when passing through two slits. This de facto leads to a wave–particle duality paradox, whereby two contradictory pictures fully explain how light and matter behave but only when these pictures are taken *together*.

Heisenberg matrices and Schrödinger's wave equation In 1925, the young physicist Werner Heisenberg was struggling to make sense of the intensities of the spectral lines when an electron jumps from one energy level to another. While stationary around the nucleus, as Bohr had shown, electrons do not radiate any energy, yet they can instantaneously jump on a higher or lower orbit whenever they absorb or emit a quantum of energy. How they did that, however, was unclear, since in these orbital jumps, the electron seemed to disappear. To tackle this issue, Heisenberg opted to concentrate on the only observables he had, that is, the frequencies of the spectral lines. And from there, he managed—with the subsequent validation by Max Born and Wolfgang Pauli—to elaborate matrices for all the possible jumps the electron could perform, eventually developing "a method to determine quantum-mechanical data using relations between observable quantities" (Heisenberg 1925, 276). It was the first mathematically consistent formalization of quantum mechanics.

One year later, building on de Broglie's work, physicist Erwin Schrödinger formalized in a wave equation the deterministic behavior (in time) of a particle. His wave mechanics provided physicists with a "familiar and reassuring alternative to Heisenberg's highly abstract formulation" (Kumar 2008, 182). Although the two solutions were later found equivalent by Paul Dirac, a heated debate between Heisenberg and Schrödinger ensued, with the former labeling Schrödinger's ideas as "crap" and the latter replying, "I can't imagine that an electron hops about like a flea." The core point of contention remained: the wave–particle duality. While both solutions could be efficaciously applied, their interpretation was in question, especially considering that Schrödinger's equation did not refer to anything directly observable, as it contained a complex number that made the wave a virtual (probabilistic) wave. It is no surprise that from there, over the decades, a variety of theories were advanced taking different paths (e.g., Bohm 1952; Everett 1957; Wigner 1961; Ghirardi et al. 1986; Rovelli 1996).

Uncertainty principle (or indeterminacy) Proposed in 1927 by Heisenberg, this principle states that it is not possible to measure *simultaneously* and with *maximum precision* two properties of particles, such as position and momentum. Experimenters can either know precisely one of the properties at each time or both of them only approximately. Reflecting on the wave–particle duality, Heisenberg was trying to solve the conundrum by foregrounding the particle side of the duality. Again, he reasoned starting from what he could observe: "We had always said so glibly that the path of the electron in the cloud chamber could be observed. . . . But perhaps what we really observed was something much less. Perhaps we merely saw a series of discrete and ill-defined spots through which the electron had passed.

In fact, all we do see in the cloud chamber are individual water droplets which must certainly be much larger than the electron"[5] (Heisenberg 1971, 77–78). In fact, for Heisenberg, there is no particle with a well-defined position or a well-defined momentum before measurement: path has no extant meaning; there is only measurement-based position or momentum, and it is the precise measurement of one of the two—as an unavoidable disturbance of the particle—that makes the precise measurement of the other impossible. This was mathematically formalized in the equation: $XP - PX = i\hbar$, where X stands for the position of a particle, P stands for its speed multiplied by its mass, i is the imaginary number $\sqrt{-1}$, and $\hbar$ is Planck's constant divided by 2π. The critical point is that the product is not commutable: $XP - PX$ does not equal 0, and this is so because here position and speed are not numbers but matrices. This uncertainty does not depend on the precision of the chosen equipment; it is a fundamental feature of reality.

Collapse of the wave function   It is the process, part of the Copenhagen interpretation, that sews together Heisenberg's uncertainty principle with Schrödinger's wave equation, in the probabilistic interpretation provided by Born. While, prior to observation, a particle only exists in the probabilistic field described by the wave function, it is only when a measurement is conducted that the particle is detected in one of its possible actual states, making all the others collapse. This phenomenon and its nature remain one of the most contentious aspects in quantum mechanics (cf. Gao 2018).

Complementarity principle (and the issue of observation)   Proposed by Bohr in 1928, the complementarity principle accommodates the idea that quantum systems allow for *mutually exclusive* descriptions (e.g., particle and wave) based on the performed experiment. Bohr (1949, 210) explains this as follows: "Evidence obtained under different conditions cannot be comprehended within a single picture but must be regarded as complementary in the sense that only the totality of the phenomena exhausts the possible information about the objects." The multiplication of experiments on the wave–particle duality urged physicists to question experimental practice itself. In other words, what is meant by measurement? How should physicists understand its effects? In the early years of the nascent quantum mechanics, physicists were very much aware of the deep philosophical implications that the new discoveries about the subatomic world brought to the fore, to the point that scientists were afraid that the whole understanding of reality that physics had developed up to that point might shatter into pieces. Bohr, who had a background in philosophy, advanced an interpretation that, while regarding the act of observation as pivotal in shaping an observer–observed phenomenon of the subatomic world, safeguarded the unambiguous classicality of its results. His words are telling:

> A sentence like "we cannot know both the momentum and the position of an atomic object" raises at once questions as to the physical reality of two such attributes of the object, which can be answered only by referring to the conditions for the unambiguous use of space-time concepts, on the one hand, and dynamical conservation laws, on the other hand. While the combination of these concepts into a single picture of a causal chain of events is the essence of classical mechanics, room for regularities beyond the grasp of such a description is just afforded by the circumstance that the study of the complementary phenomena demands mutually exclusive experimental arrangements. (Bohr 1987, 40–41)

This means that, on the one hand, at stake is not only which observables one can(not) know at any given time but also the recognition that the whole act of observation determines the phenomenon one can observe (and, in the case of the wave–particle duality, these are mutually exclusive phenomena): the measuring apparatus and the system being observed are entangled; on the other hand, the final interpretation of the experiment remains confined within a classical frame through the unambiguous use of classical concepts.[6]

Superposition   In a quantum system, superposition is the phenomenon whereby particles are found, probabilistically, in overlapping physical states, such as spin up and spin down. Once the particles are measured, then superposition disappears, and their state takes a definite value. When speaking of particle states, it is worth retracing how terms such as *spin* and *rotation* entered the language toolkit of quantum mechanics, showing the attempt to iteratively fine-tune scientists' vocabulary in parallel with the deeper and deeper discoveries of the subatomic world that did not match phenomenological reality. A passage from Kumar's (2008, 152) book is telling in this respect:

> An electron could move up and down, back and forth, and side to side. Each of these different ways of moving physicists called a "degree of freedom." Since each quantum number corresponds to a degree of freedom of the electron, Uhlenbeck

believed that Pauli's new quantum number must mean that the electron had an additional degree of freedom. To Uhlenbeck, a fourth quantum number implied that the electron must be rotating. However, spin in classical physics is a rotational motion in three dimensions. So, if electrons spin in the same way, like the Earth about its axis, there was no need for a fourth number. Pauli argued that his new quantum number referred to something "which cannot be described from the classical point of view." . . . Goudsmit and Uhlenbeck provided the first concrete evidence that existing quantum theory had reached the limits of its applicability. Theorists could no longer use classical physics to gain a foothold before "quantising" a piece of existing physics, because there was no classical counterpart to the quantum concept of electron spin.

Superposition can be rightly considered—from a classical point of view—one of the main counterintuitive behaviors of the subatomic world in that it posits the necessity to think beyond (classical) phenomenology and in terms of probability.

Entanglement One even "stranger" phenomenon than superposition is entanglement. Two particles are entangled if the measurement of one of them (e.g., its position or momentum) instantly affects the other particle, irrespective of any distance. In other words, the states of entangled particles cannot be measured independently from each other. This means that entanglement introduces a *simultaneous codependency* that breaks away with classic cause–effect relations and, in particular, the "locality"—an object is directly influenced only by its immediate surroundings—and "realism"—objects exist independently of observation—of the laws of classical physics.

This was at the core of the fundamental disagreement between Bohr and Albert Einstein on how to make sense of quantum physics. Einstein subscribed to local realism entailing that reality exists independently of observation, while Bohr endorsed a vision in which physical reality is not "populated" by given individual objects with preexisting properties; rather, these objects get determined depending upon observation.

In 1935, Einstein, Boris Podolsky, and Nathan Rosen (1935, 777) wrote an article in which they tried to disprove entanglement: "If, without in any way disturbing a system, we can predict with certainty the value of a physical quantity, then there exists an element of reality corresponding to that quantity." What all three supported was a realist position according to which, given entangled particles, these have properties that preexist the measurement. Hence, it cannot be that the measurement of one of them simultaneously determines the other: either quantum physics violates local realism, or the theory is incomplete, meaning that, at a deeper level, there must be hidden variables that the theory does not take into account (this has come to be known as the "EPR paradox"). It is only in 1964 that John Bell demonstrated that Einstein, Podolsky, and Rosen were wrong: in fact, he proved that not only the EPR paradox but also all local theories, including hidden variables, are incorrect (what is known as Bell's inequality theorem). From Bohr's (1935, 700) perspective, even before Bell's demonstration, to be at stake in the EPR paradox is an ontological issue, notably a wrong conception of what is real:

> The wording of the above-mentioned criterion of physical reality proposed by Einstein, Podolsky, and Rosen contains an ambiguity as regards the meaning of the expression "without in any way disturbing a system." . . . There is essentially the question of an influence on the very conditions which define the possible types of predictions regarding the future behaviour of the system. Since these conditions constitute an inherent element of the description of any phenomenon to which the term "physical reality" can be properly attached, we see that the argumentation of the mentioned authors does not justify their conclusion that quantum-mechanical description is essentially incomplete.

While for Einstein, Podolsky, and Rosen, the issue of "disturbing the system" was crucial, for Bohr, observer and observed are not separated realms but part of the same whole; hence, they necessarily mutually determine each other. Put differently, for Bohr, the EPR paradox rests on the wrong interpretation of what quantum physics says (or can say) about the nature of reality.

De/coherence *Coherence* is the state of a quantum system fully isolated from the environment, allowing phenomena such as superposition and entanglement. As long as a quantum system is isolated, it is coherent, and its particles behave quantum mechanically. When a quantum system is not perfectly isolated, it undergoes a process of *decoherence* and its particles no longer show quantum mechanical behaviors. This is so because random interactions intervene, and as a consequence, wave interferences in the system become randomized, dictating particles to behave classically. However, decoherence is neither a one-way process, insofar as it is reversible on a very, very long time frame (cf. Smolin 2019), nor is it a process that occurs at a certain threshold along the micro–macro axis of physical reality, as Irina Basieva, Andrei Khrennikov, and Masanao Ozava (2021) advance with regard to the quantum-like modeling of open information biophysical systems. Decoherence is a contingent matter; it depends on the scale and time frame one adopts to identify and arrange an isolated system.

Quantum field theory (QFT) In physics, a field is a region of space in which each point is affected by a force. Today, scientists regard fundamental subatomic particles (fermions)—up quark and down quark (which make up protons and neutrons), electrons, and neutrinos (known as leptons, the former with charge, the latter without charge)—not as "entities," in the traditional sense of the term, but as bundlings that "pop up" from their own underlying field. This "popping up"—better called excitement—is due to the interaction among particles' underlying fields through four force fields (whose corresponding particles are called bosons): electromagnetism, the strong force (that keeps quarks together), the weak force (involved, for instance, in radioactive decay), and gravity (which, according to Einstein's general relativity, is space-time itself). To this, it must be added the Higgs field (whose particle was empirically observed in 2012 at CERN), responsible for the interaction among particles. Today, the so-called Standard Model—the most complete picture of physical reality humanity has got so far—includes twelve particle fields (the four particles above are repeated twice over with different masses), the four force fields, and the Higgs field.[7] This model can also be thought of as a quantum theory but not one that can simply be achieved by quantizing physical quantities with corresponding operators; rather, it is a field theory where it is fields that are quantized.

QFT is a local theory combining the "discreteness" of energy and matter with the notion of field. So far, QFT represents one of the quantum-based best descriptions of the subatomic world (note, however, that QFT, like the Standard Model, has not been able so far to include the field of gravity into its model). Computer simulations have helped visualize what happens in the "vacuum," showing the "restlessness" of the fields that make up particles *even* in their absence, that is, when these fields are at lowest energy states. In 2023, a team of research (Roques-Carmes et al. 2023) at the Massachusetts Institute of Technology (MIT) managed for the first time to control the quantum randomness of vacuum fluctuations, opening new prospects for quantum-enhanced probabilistic computing.

*   *   *

To make sense of the oddities of quantum physics, it is necessary to reconsider what (we think) we know about the nature of reality, how we claim to know it, and, more broadly, what we think reality is and how we are enmeshed with it. Clearly, philosophical issues are at stake here, but physics, after all, is—has always been—about testing and being tested by philosophy. In *The Quantum Challenge*, George Greenstein and Arthur Zajonc (1997, 144) claim that quantum physics' "metaphysical implications are profound. [They] go so far as to change the very way we should think of physical existence at its most fundamental level." And yet, holding on to reality-as-we-know-it is tempting and reassuring: as Nobel Prize winner Anton Zeilinger (1996) notes, refusing to engage with the philosophy of quantum physics "often is motivated by trying to evade its radical consequences, that is, an act of cognitive repression on the part of the proposers." Continuing to do so, however, might have undesired consequences: "Where did we ever get the strange idea that nature—as opposed to culture—is a-historical and timeless?" Steven Shaviro (1997, 39) provocatively asks in *Doom Patrols: A Theoretical Fiction about Postmodernism*. Nature changes and it changes humans from within; or better, nature *is* change, and people are part of it, for better or worse. According to Evelyn Keller (1985), "What is required is a paradigm that on the one hand acknowledges the inevitable interaction between knower and known, and on the other hand respects the equally inevitable gap between theory and phenomena." Indeed, what quantum mechanics implies, at its core and beyond any disciplinary boundary, is that ontological, epistemological, and ethical concerns are deeply intertwined in shaping reality, and these require new conceptual frameworks and tools to be tackled conjointly.

Laying the Ground: Defining Concepts, Unpacking the Argument

Equipped now with an understanding of the main ideas and debates of/ around quantum mechanics, here we provide other terminological clarifications that will complete our conceptual toolbox. In doing so, we will be able to further unpack our argument explaining how we plan to proceed.

Realities, Embodiment, Worldviews (or "World-Sensing")

Based on the discussion in the preceding section, it follows that the very notion of reality is under scrutiny. Physicist John Wheeler (1979) once wittily stated that "no word is a word, until that word is promoted to reality by the choice of questions asked and answers given." This is a brilliant idea because, while stressing the emergence of reality as a codependency of questions and answers (another way to see the observer–observed dualism), it unveils the arbitrariness of any linguistic expression until it is anchored to a well-defined context—and it does so by speaking precisely of reality.

Here, far from espousing an essentialist idea, we regard and treat reality as a shared (sociotechnical) process. Reality must be intended, at all times, as constructed in the very act of observation or—as we call it, to avoid the risk of being confined within representationalism—*embodiment*. Embodiment is a fit(ter) term for two reasons. On the one hand, it can be extended not only to animals and plants but also to all organisms able to seek, process, arrange, and pass on information—that is, *autonomous organisms*—and we would not be surprised to see this extension reaching out soon to artificial entities. This is an *autopoietic* understanding of embodiment: *autopoiesis* is a term coined by Humberto Maturana and Francisco Varela (1980, 78) to indicate the ability of a system to organize as a unity and remain so over time (homeostasis), that is, "a network of processes of production (transformation and destruction) of components which: (i) through their interactions and transformations continuously regenerate and realize the network of processes (relations) that produced them; and (ii) constitute it [the embodiment] as a concrete unity in space in which they (the components) exist by specifying the topological domain of its realization as such a network." To give an example, this is de facto how the DNA works: in order to have self-sustaining life, a recursive process of reproduction and differentiation—what Douglas Hofstadter (1979) calls "strange loops"—is necessary. In order to do this, the biological system (the primordial entity) needs to contain in itself (1) a set of information (genes) to be replicated, (2) an interpretative procedure to have that information replicated, and (3) a replicating-differentiating procedure of the information to be copied. As Carlos Gershenson (2015, 866) puts it, "Autopoiesis considers systems as self-producing not in terms of their physical components, but in terms of its organization, which can be measured in terms of information and complexity." In this respect, embodiment shall be regarded as a fractal concept: far from equaling the "body," it is rather the action that counts, pointing to an endless coupling of individual and collective dimensions—as complex systems—in an implicated order of mutual dependency.

On the other hand, embodiment helps bypass the possibly misleading idea of observation as a necessarily conscious practice: as Aaronson (2013, 164) acutely writes, "When you consider all the junk that might be entangling itself with your delicate experiment . . . it's as if the entire rest of the universe is constantly trying to 'measure' your quantum state." The idea of embodiment, then, de-anthropomorphizes and de-instrumentalizes observation and measurement, positing, instead, that the act of observing/ measuring is consubstantial to the coming into being of physical reality, regardless of whether an experiment or a given technological setup is arranged. This means that any living organism is always already in a condition of "observership"; such an organism is therefore *technological* for the very fact of living. Expanding on the notion of "enactment," which emphasizes that "cognition is not the representation of a pre-given world by a pre-given mind but is rather the enactment of a world and a mind on the basis of a history of the variety of actions that a being in the world performs" (Varela et al. 1991, 9), what we claim here is that living is in itself an ongoing instantiation of "observership" intended as a coevolution between self and the world (i.e., embodiment).

Along the same line, we find the term *worldview* somewhat inappropriate. Indeed, worldview retains a chiefly visual connotation assuming a static and unique point of view (cf. Haraway 1988). By contrast, to the extent to which quantum mechanics questions the very idea of a world out there to be known objectively and regards, instead, experience as an emerging phenomenon connecting observer and observed, we prefer to adopt the term *world-sensing*. This term has some advantages: (1) it is not attached to any specific organ; (2) it contains the idea of "sense," as a word with a threefold connotation: *direction, feeling*, and *meaning*, thus encompassing both the perceptual and the cognitive (cf. chapter 2); and (3) the term is in the "ing" form because at stake is, at all times, a process, not the reification of experience (the best one can do is to become aware of some moments of this process whenever one self-reflects on its "being in the world").

This, however, does not imply to embrace a flat relativism: within physical reality, more or less stable onto-epistemological formations can be detected, based on certain sociotechnical enframing—or ecologies. The notion of reality, then, acquires meaningfulness from within or by comparing ecologies. On this, we embrace Thomas Kuhn's (1970) perspective arguing in favor of the (partial) onto-epistemological incommensurability of different ecologies, based on a mix of perceptual-observational and semantic factors. Simply put, the fabrication of reality each ecology produces is unique as it is based on a specific sociotechnical paradigm. However, at no time does an ecology exist independently from others, nor does their partial incommensurability prevent comparison. In fact, we are particularly interested in exploring how ecologies intersect and the effects they produce.

To launch such a project becomes urgent at a time of deep technological transformation, environmental threats, and societal instabilities. What Bernard Stiegler (2017, 136) calls "organological life," that is, the intertwining of the individual with technological artifacts and social systems as a whole, is based on "*epokhal* technological shocks" that require ongoing readjustments of the entire organological assemblage. This shock is the historical-material substantiation of what new technologies embed with regard to the milieu they appear in and by which they are informed. The emergence of the internet, for instance, has allowed the digital transformation of the present, disrupting the language ecology and reconfiguring people's experience of/in the world: "All [these] developments," Elena Esposito (2017, 287) writes, "call into question the simple distinction between the computer and its environment and between machine and user." Quantum technologies will further build upon that, and they will do so by concretizing the onto-epistemological tenets of quantum mechanics through a radically different enframing.

To be clear, then, although we will sometimes resort to the term *reality* for the sake of simplicity, whenever we will refer to the specific reality produced by/within/through a given ecology, we will adopt the following notation: "reality" for the language ecology, /reality/ for the digital ecology, and {reality} for the quantum ecology.

Ecologies

The term *ecology* was first coined in 1866 by zoologist Ernst Haeckel, who used the term *oekologie* to describe the "relation of the organism to the organic as well as its inorganic surrounding environment." The word comes from the Greek *oikos*, meaning "home" or "place to live," and *logos*, which means "discourse" or "study." The purposely broad conceptualization of *organism* and its relationship with the environment, as both organic and inorganic, is essential here. This conveys the idea of an immersion—a symbiosis—between living and inert elements, and it considers the surrounding as a pulsing dimension. It also follows that in this book, *ecology* does not stand for "environment" only, nor does it relate exclusively to climate issues; rather, it encompasses a holistic vision.

Recently, ecology has been associated with various fields—animal ecology, human ecology, agricultural ecology, and so on—and it has also been extended to the media. Marshall McLuhan introduced the idea that media are shapers of the (sociocultural) environment of which they are part: "Media ecology looks into how media affect human perception, understanding, feeling, and value," media theorist Neil Postman (1970, 161) notes. It is along this line that we understand "quantum ecology," that is, how quantum technologies will shape {reality} both epistemologically and ontologically beyond a classical conception and apprehension of the world. Notably, we aim to do so by reconceptualizing "technology," "environment," and "human" as thick aggregative processes (rather than entities), whose imbrication is constitutive of the new ecology. In fact, more than a triad and a set of relations, an ecological approach de-essentializes these terms and their links and conceives them beyond mere recombination, in view of a synthetic understanding of reality in which anthropological, technological, and environmental dimensions are consubstantial.

On this point, Alexander Wendt's (2015, 229) reference to holograms helps elucidate the idea: "Holograms are holistic in the sense that all of the information about the object is recorded on each pixel of the film. . . . The whole is present *in* the parts, not made up *of* them." This is a good characterization of the quantum ecology: individuals, technologies, and environments are not given, distinct, and separate (f)actors once and for all; rather, each of them *is also already* the others, so that their coming into being is a reiterative emergent practice that constantly does and undoes boundaries.

Sewing together these reflections, cosmologist Thomas Hertog (2023, 174) writes that "strikingly, quantum entanglement appears to be the central engine that generates gravity and curved space-time in *holographic physics* [italics added]. It is what laser light is for an ordinary optical hologram."[8] Hence, according to him, entanglement sits at the core of an information theory-based formalization of physical reality. From our perspective, if entanglement is the "engine" of the quantum ecology, uncertainty and discreteness are its structural dynamic components, at once cause and effect of a holographic order of existence (cf. chapter 5).

*   *   *

In this book, we identify three ecologies, each characterized by a trademarking *dispositif*: language, digital, and quantum. Each *dispositif* enacts its own *operating systems*, which then find applications in context, giving rise to consistent (but not necessarily cohesive) *technocultural fields*. These fields come to life as an interlaced mix of *actors*, both physical and legal (e.g., human and nonhuman); *factors*, both tangible and intangible (e.g., tools, competences, skills, economic and natural resources); and *discourses*, from policy documents to media contents, ideas, and ideologies. Our phenomenological approach throughout the book will be to unpack these ecologies along three dimensions—*space, time,* and *identity*—showing how actors, factors, and discourses materialize certain onto-epistemological formations across the world.

Language ecology Language is the *qualifying dispositif* of human communication. The key thing is that language—be it oral or written—has developed in order to produce *meaning*, which is the principal bond among people (i.e., a sociotechnical shared construct), allowing one to make sense of the environment as well as one's own experience. In fact, two key principles of language are (a) to *signify* the world and communicate it and (b) to be *autopoietic*, that is, language is able to refer to itself and get recursively disentangled from within thus adapting and evolving. Language is here further articulated in enactive (at once, representational and operational) operating systems—alphabetic and logographic—each one giving shape to its own technocultural fields. The former is where applications like selfhood and classical physics, just to provide two examples, find their roots; the latter favors holistic thinking and a collective-based world-sensing. As Yuval Harari reminds us, language has been the principal operating system of humanity:

> Over the past couple of years new AI tools have emerged that threaten the survival of human civilisation from an unexpected direction. AI has gained some remarkable abilities to manipulate and generate language, whether with words, sounds or images. AI has thereby hacked the operating system of our civilisation.[9]

This is a good exemplification of what we have visualized in figure I.1, where we show the (partial) complementarity between the *dispositifs*, operating systems, and effects of the language and digital ecologies. Vertical arrows are discontinuous because they are not given but repeatedly actualized, and they are bidirectional because levels mutually affect each other. Technocultural fields are signaled by green dotted rectangles. In chapter 2, we will

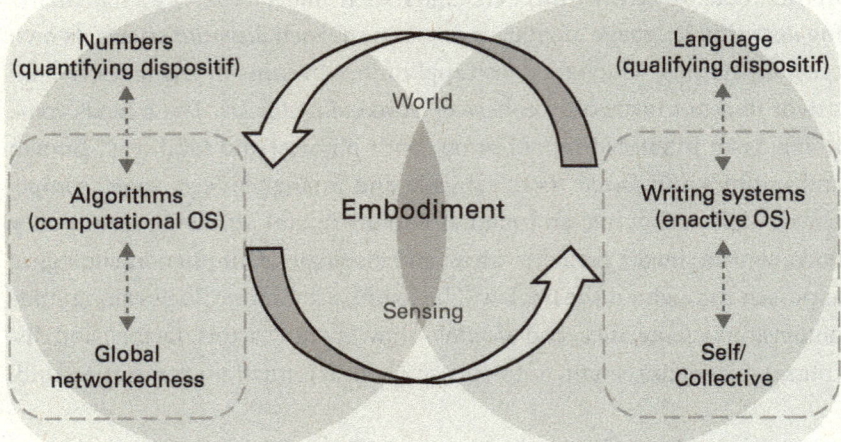

Figure I.1
Language and digital ecologies.

explore at length the epistemological tensions running through the different world-sensing created by these two ecologies.

Digital ecology   At the core of the digital ecology is a *quantifying dispositif*: numbers. Algorithms are the *computational operating system* at the basis of the digital transformation. Algorithmic computation is not predicated on meaning but on the *formatting* of physical reality into binary code. Two key principles of the digital ecology are (a) the turning of information into bits to be *processed* and (b) performativity, that is, the *isomorphic* processing of such "datafied" information (all data are treated equally). This, in turn, produces an epistemological convergence of technocultural fields in the form of *global networkedness*. As Franco Berardi (2015, 21) notes, through digital technologies, the singularity of the subject as "a way of becoming" gets progressively reduced to "a set of components, or a format." Key applications of the digital ecology are, among others, the datafield, data regimes, digital twins, and the metaverse.

Quantum ecology   It is based on quantum mechanics (while the other two ecologies emerge from a classical understanding of physical reality. In fact, quantum information technologies engineer computation by manipulating particles, not classical bits). More precisely, quantum ecology rests on three key tenets: *uncertainty*, *discreteness*, and *entanglement*. Aaronson (2013,

110) proposes to look at quantum mechanics itself as an operating system, since it "sits at a level *between* math and physics that I don't know a good name for. Basically, *quantum mechanics is the operating system that other physical theories run on as application software*. There's even a word for taking a physical theory and porting it to this OS: 'to quantize.'"

There are two main reasons for considering quantum as an emerging ecology: (a) it will host the next generation of core information technologies, a feature it shares with the other two ecologies also responsible for technological innovations, from the press to mass media (language ecology), to the internet and data-driven solutions (digital ecology), and (b) it forces one to reconsider the nature of reality itself, based upon quantum physics' onto-epistemological tenets, thus also holding the key to changing the perception of (and action upon) the other two ecologies.

The operational logic (in our case, operating system) of the quantum ecology is *synthetic*: the etymology of *synthesis* is "putting together, composition"; it implies the (re)combination of parts—discretization—as well as the creation of something that is not necessarily the sum of these parts but a new whole (where parts are codependent), and in this process, a qualitative change occurs that is unpredictable (i.e., it cannot be predetermined). At the same time, synthesis relates to something artificial, not natural, or also, simulated, inevitably transfixing epistemological and ontological boundaries. The quantum ecology will produce *ultra-experiential* effects, beyond the gnoseological and phenomenological dimensions of human experience, yet the shape it will take will depend on how its actors, factors, and discourses will play out. Possible applications will go from competing geopolitical bubbles and all-immersive simulations to a reconsideration of communication itself. Figure I.2 visualizes the emerging quantum ecology: again, lines are discontinuous because the way in which the synthetic operating system will operationalize its principles will be in the fashion of an emerging dynamic, not a static affair. The investigation of the overlap between quantum technologies and embodiment, as well as how this will affect both language-based and digital-based applications, will be at the core of chapter 5 (it is not possible right now to identify technocultural fields, but we will provide some suggestions).

Sociotechnical Paradigm

Sociotechnical paradigm is a term used here as synonym of ecology, although this latter term remains preferable in that it brings to the fore the synergic

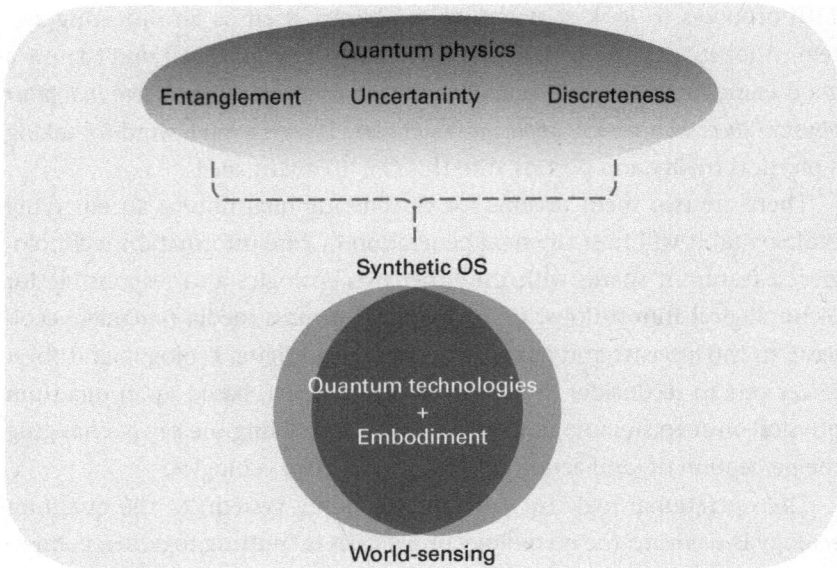

Figure I.2
The quantum ecology.

coevolution of people–environment–technology in the shaping of unique world-sensing. More precisely, the sociotechnical paradigm stresses the idiosyncratic codependency of society and technology, through interlaced actors, factors, and discourses. While often used in the singular, by no means, through this term, do we want to conceive of society and technology as homogenizing concepts. Quite the contrary, we will embark on a journey across different regions of the world, as well as focusing on different sets of technologies (in the plural), with the goal to highlight dis/continuities across technocultural fields (with culture as the collective instantiation of embodiment).

Technologies   As mentioned, we understand technology in a Foucauldian sense, that is, as a *dispositif* embracing "a thoroughly heterogeneous ensemble consisting of discourses, institutions, architectural forms, regulatory decisions, laws, administrative measures, scientific statements, philosophical, moral and philanthropic propositions" (Foucault 1980, 194). Building on Foucault's work, Giorgio Agamben (2009, 17) extends such a concept to "literally anything that has in some way the capacity to capture, orient, determine, intercept, model, control or secure the gestures, behaviors, opinions or discourses of living beings." This characterization has the merit to put the stress on processes rather than reified instantiations. One could contest that, according to Agamben, everything gets to be regarded as a *dispositif*, but this is really to miss the point. At stake is rather the acknowledgment of the fabrication—the ongoing *mis-en-forme*—of reality under certain bio-sociotechnical conditions. What is interesting to us is that the French term *dispositif* comes from the Latin *dispositio*, which means "organization, arrangement." Hence, the idea of technology as *dispositif* is particularly fitting because it immediately aligns to that of enframing, allowing to unpack how certain setups—of actors, factors, and discourses—within a given ecology inform the fabrication of a precise world-sensing. As Karen Barad (2007, 105) notes, "Apparatuses are specific material reconfiguring of the world that do not merely emerge in time but iteratively reconfigure space-time-matter as part of the ongoing dynamism of becoming." This means that, past the idea of a loop connecting human, technology, and environment, it is key to regard this triad as an emerging whole that repeatedly redefines its own onto-epistemological horizon.

Concerning language, we do not favor any privileged point of reference, resting on the premise that, since language is a qualifying (organic) *dispositif* meant to produce meaning, it is always already shared, and language immerses the individual into the context as much as it comes from it. In this regard, Charles Taylor (1986, 133) notes that "the linguistic capacity is essentially more than an intellectual one; it is embodied . . . [it] is essentially shared: it sustains a shared consciousness of the world." This and similar positions, while reasserting the embedded nature of language, tend to extend its "validity" to humankind as a species, mapping an isomorphism between language and body as *dispositifs*. More phenomenological approaches, such as that by Maturana and Varela (1987), regard language as a milieu for self-reflection that enables humans to perceive and communicate their own presence and situatedness. For these scholars, then, languages coevolve with the body, which becomes the true monad of human experience, working as a form of inscription of human beings in the world. What is worth stressing, beyond these differences, is that language does not convey meaning as a given that preexists experience but fosters signification as an embodied and culturally embedded process.

With regard to the digital ecology, technology broadly refers to information and communication technology (ICT) infrastructures—as the backbone that since the 1970s has allowed for telecommunications with a global outreach to emerge—and data-driven technologies, that is, the sets of algorithmic apparatuses that, by capitalizing on ICTs, exploit digitalization and deep learning for providing increasingly responsive and customized services. Having said this, technology does not, alone, bring about change. To the extent data-driven technologies are pervasive in everyday life, they must be investigated in conjunction with other socioeconomic-political processes. Studies have warned about data's imbrications with power relations, social (in)justice, and cultural biases (Dencik et al. 2016; Kitchin 2014; Metcalf and Crawford 2016). Efficiency and customization of services are just two facets of the techno-rationalization of society via data-driven solutions; other facets, such as the commodification of the individual, the risk of reinforced and new forms of discrimination, exclusion, and inequalities, or even the emergence of surveillance regimes, stand as daunting countereffects (Brannon 2017; Eubanks 2018; Hintz et al. 2017).

As for the quantum ecology, Jonathan Dowling and Gerard Milburn (2003) claim that humanity has entered the second quantum revolution: the first one concerned the discovery of the behaviors and principles of quantum mechanics; the second one has to do with how these behaviors and principles are put to use. Emerging quantum technologies are a set of groundbreaking innovations that will have a *systemic societal impact* (WRR 2021). They will, at once, refine and disrupt today's digital transformation and the language ecology, redesigning the way to make sense, re-create, and synthesize reality, by realizing at scale the onto-epistemological tenets of quantum physics.

More specifically, it is possible to distinguish between "quantum sensing technologies" (e.g., quantum metrology, quantum navigation, and quantum imaging) and "quantum information technologies" (QITs) (e.g., quantum computation, quantum simulation, and quantum communication). In the former category fall technologies already in use today: for instance, the increasing accuracy of Global Positioning System (GPS) exploits the entanglement effect, while magnetic resonance imagining (MRI) for medical diagnosis exploits the spin of the hydrogen's proton. The second category relates more properly to those macro technologies whose advent—although still in the making[10]—will have a pervasive impact throughout society. It is on these that we will concentrate our attention.

Societies Our point throughout the book is that accounting for societal and cultural differences is decisive: to support a global order that is ecological, one needs to look at societies not as given but as multifaceted dimensions traversed by internal tensions as much as mutual interdependence. Beyond networks (Latour 1987), what is being configured are complex heterogeneous "wholes": "social totality," Aníbal Quijano (2007, 177) correctly notes, "requires the idea of an 'other' that is diverse, different" and that is immanent (not heterotopic) to that same totality. To consider the world as a complex heterogeneous whole, then, entails to foreground that processes determine phenomena (not the other way around), and neither of the two can be studied in isolation or frozen in an atemporal snapshot. The coming into being of the whole is something more than *and* different from the sum of its constituents.

Concerning the language ecology, we mostly focus on alphabetic and logographic operating systems. This allows us to look at different cultures in different regions of the world—in the West and the East—and to connect them, cutting through geographical and political borders. In this sense, our rationale will follow as much an international as a transnational logic.

As far as the digital ecology is concerned, while it is true that it is in North America and Europe that the digital transformation has taken the lead since the 1990s, by now this process has become global as well as trans-sectoral—involving public, private, semipublic, institutional, and noninstitutional actors—thus being shaped by contextual factors that determine its ongoing coming into being. In this regard, we strive to bring into the discussion diverse examples, connecting societies and technological innovations from across the globe, not only along the West–East–North–South axes but also within each technocultural conglomeration.

The same rationale will also guide the investigation of the quantum ecology. China plays an increasingly relevant role—if not one of primacy—in the field of quantum technologies, with reverberations that spread across the world, including southern regions. At the same time, other actors—especially private companies—will have a say in the shaping of the whole ecology, creating their own field of operations that might align and/or conflict with that of major (geopolitical) powers. It is these tensions that we will try to unpack.

Quantum Physics in the Social Realm: Mapping the Scene

Over the past thirty years, there has been a resurgence of interest in quantum physics not only due to its increasing number of applications but also in conjunction with its onto-epistemological meaning and what this ethically entails. This has led to a variety of works that, while interconnected, differ in some methodological respects.

In the first group fall those studies and scholars that adopt a "quantum-like approach" in the attempt to quantize social phenomena. In physics parlance, quantum-like is, more than a figure of speech, a method of analysis, and a description of a target—for example, a concept, social situation, behavior, or form of cognition that borrows phenomena well established in quantum mechanics to help better understand the object under examination. In this respect, the principles and formalism behind quantum mechanics not only are a lens through which to scrutinize individual and collective phenomena but also become a predictive modeling that goes beyond what classical paradigms can tell. This can lead to rethinking hitherto classically approached issues and even entire fields in quantum terms. Examples abound in areas as diverse as economy (e.g., Haven and Khrennikov 2013; Ikeda and Aoki 2022), linguistics (e.g., Aerts et al. 2006), psychology and cognition (e.g., Gabora et al. 2008; Kauffman and Roli 2022), decision theory (e.g., Busemeyer and Bruza 2012; Aerts et al. 2019), neuroscience (e.g., Penrose 1994; Shimony 1997), and biology (e.g., Abbott et al. 2008; Ball 2011; Al-Khalili and McFadden 2015; Atmanspacher 2017). Given our background rooted in the humanities and social sciences, we avoid venturing in the modeling of our ideas according to the math formalism at the basis of quantum mechanics. Nonetheless, we do acknowledge that such a "quantum-like" approach is fruitful for bridging physics and social realms. In fact, "quantum-like" and "ecology" can be regarded as the two sides of the same coin. More precisely, while quantum-like scholars—in force of their background—tend to start from quantum mechanics to move toward the social realm, we move from the social realm toward quantum mechanics, focusing more on the ecological side of the story (i.e., a self-organizing codependent dimension linking human, technology, and environment).

In the second group are works that do not resort directly to quantum mechanics' formalism, and yet, they question their own object or discipline through quantum physics. As our ecological approach entails, the present book aligns more closely to this second group, although it does incorporate findings also from the first group. Notably, we expand upon existing works by scholars such as, but not limited to, David Bohm ([1980] 2002), Danah Zohar and Ian Marshall (1994), Karen Barad (2007), Alexander Wendt (2015),

Jennifer Burwell (2018), and Timothy Eastman (2020). From/blending different perspectives (e.g., cultural studies, philosophy, literary criticism, sociology, international relations), all these works share the same urgency to rethink reality-as-we-know-it and, relying on the teachings of quantum mechanics, question basic disciplinary assumptions on language, the nature of reality, knowledge, and societal issues, providing, in turn, provocative insights into the fabrication of what is assumed to be given and fixed, and enabling a deconstruction/critique of its very fabrication.

Overall, we feel sympathetic toward these works, especially as they attempt to connect realms that would otherwise remain isolated. In fact, we took great inspiration from the authors mentioned above, and we made sure to embed their major lessons in this book. We are also aware, though, that these—and similar—works are sometimes subjected to harsh criticism coming mostly from those orthodox scholars—from all sides—who strive to defend their own disciplinary horizon—with good rights, probably, but not always with good reasons. After all, one teaching of quantum mechanics is that it is hard—if not impossible—to trace neat boundaries, especially when it comes to knowledge and knowing. Overall, we believe that, as long as research and argumentations are conducted rigorously and fairly, conceptual and methodological hybridization is—should be—the rule, rather than the exception. As Heisenberg (1971, 70) noted, "In science, too, it is impossible to open up new territory unless one is prepared to leave the safe anchorage of established doctrine and run the risk of a hazardous leap forward. . . . When it comes to entering new territory, the very structure of scientific thought may have to be changed, and that is far more than most men are prepared to do." We take this as a call to action.

The present book speaks from the vantage point of (and delves into) the boom of quantum information technologies, which have entered political agendas worldwide. While it is true that technologies of common use exploiting quantum phenomena have been around for decades, it was only in 2019 that Google declared to have reached quantum supremacy with its Sycamore computer (cf. chapter 4). Of course, this was just a milestone: since then, research has continued, supported by heavy public–private investments around the world, especially in the United States, Europe, and China, and it is not difficult to expect further developments and applications of quantum information technologies in the upcoming years. Michio Kaku's (2023) *Quantum Supremacy* is probably the most recent attempt to deliver an overview on this subject. Yet, the book remains futurologist in the treatment of technology, sidestepping a critical discussion of the upcoming paradigm. Instead, by inscribing our analysis along the line of language and digital transformation, we aim to imprint a unique twist on the subject, and we will do so, on the one hand, by focusing on the sociopsychological effects of quantum technologies on both individuals and collectives and, on the other hand, by adopting a historical-intercultural perspective that helps contextualize the analysis and explore the roots of the sociotechnical paradigm's shift.

Structure of the Book

The rationale of the book is to explore how the three ecologies have evolved, how they interact, and with which consequences: "any new technology," McLuhan (McLuhan and Zingrone 1995, 341) reminds us, "tends to create a new environment. . . . It is in the interplay between the old and new environments that there is generated an innumerable series of problems and confusions." We aim then to delve into these "problems and confusions" by unpacking the actors, factors, and discourses of each ecology and showing their mutual interdependence.

Hence, in chapter 1, we focus on the nexus body–language as crucial to unpack how meaningful world-sensing has come into being. Notably, alphabetic and logographic operating systems have fostered technocultural fields with precise features: for instance, the vision underpinning classical physics and individuality can be traced back to ancient Greeks, while in the East, reality has been long conceived as a balance (and merging) of opposites (closer to Einstein's general relativity as far as space-time is concerned) in favor of the collective over the single subject. At the same time, while traditional forms of (analogue) technologization of language—writing, the press, mass media—have overall maintained the language-based world-sensing consistent over time (although certainly not without internal frictions), today we detect signs of disruption alongside the consolidation of the internet as an early digital technology.

From here, in chapter 2, we describe how the emergence of algorithms has marked an incommensurable divide with language, a divide that is, indeed, at the heart of the *epistemological crisis* of the present. This is because language and digital ecologies are misaligned, with data-driven technologies overlapping and conflicting with language-based systems of signification. While language functions as a "senseful" meaning-making *dispositif* that provides humans with the possibility of reaching a certain level of consensus, algorithms do not mediate; they simply commute atoms into bits, enforcing mere computability. In fact, data do not "make sense" (in the anthropological sense of the term); they only make the world account-able. These two epistemological realms are incommensurable, leading to a widespread condition of absurdity, characterized—we claim—by self-sabotaging social behaviors, as the symptom of the impossibility of currently reconciling this divide.

In chapter 3, we show how, through the digital transformation led by algorithmic technologies, we witness an increasing entanglement of human life, technology, and the (responsive) environment as the ultimate realization of the digital ecology. In fact, the digital transformation has fabricated a *datafield* that is not only quantifying individuals and social phenomena but also qualitatively changing them. The effects can be articulated along three lines: (1) algorithmic technologies have foregrounded a *kairological* fabrication of time (over a chronological one), which has consequences on memory (understanding of the past) and imagination (conceptualization of the future); (2) algorithmic technologies have remodeled the conception of space, tending toward point-like ubiquity and virtualization, which means deterritorialization and ubiquitous location, with consequences on relationality and accountability; and (3) algorithmic technologies have impacted on identity, epitomized by the emergence of digital twins and multiverses, intended as digital/data doubling of individuals and physical reality, via the aggregation of globally distributed data that are then poured into increasingly entangled environments. The trends identified in chapter 3 are at the core of what we name *datacracy*, the power of data. More specifically, datacracy describes a de-subjectified system where the idea of governing through algorithms is turned into the scenario of being governed by algorithms, regardless of the technocultural field one finds oneself in (e.g., "surveillance capitalism" in the West and "social credit" systems in China). Within this scenario, various *data regimes* emerge from within the datafield as entangled bundlings that strive to keep the upper hand on the digital transformation.

In chapter 4, we take the lead from datacracy to discuss how data regimes inevitably resonate/compete with each other, especially in the context of emerging quantum geopolitical fields. Chapter 4, then, maps the strategies and discourses of major *quantum regimes* worldwide (e.g., the European Union, the United States, and China, as well as companies) with regard to the development of quantum technologies. We will argue that these latter not only are quantitative accelerators of the digital transformation and datacracy but also represent qualitative game changers. On the one hand, the advent of quantum information technologies will deeply remodel people's perception of {reality} and force them to reconsider their relations with technology and the environment. On the other hand, the emergence of quantum information technologies risks exacerbating the already consolidated multipolar geopolitical scenario, with consequences over political, social, and individual self-determination.

In chapter 5, we take the cue from quantum mechanics' basic tenets (i.e., uncertainty, discreteness, entanglement) to envision an open (communitarian) quantum ecology, aware that the concrete actualization of such ecology will depend on how quantum regimes, factors, and discourses will evolve. In the first part of the chapter, we discuss how the quantum allows to redesign both space and time: space as a relational nonlocal experience and not as an absolute dimension; time as not only a sequential experience but one augmented with a contextual appreciation of its appropriateness. Moreover, subjectivity gets reworked—following recent studies in cognitive science and neuroscience—as an already (entangled) social self. In the second part of the chapter, we discuss concrete examples on how the quantum ecology might be realized in education, economy, tech innovation, urbanism, and the arts. These examples shall not be regarded as disciplinary silos but as methodological keys refracting what it means to think of and act within the quantum ecology.

Back to the Future

In the last chapter of the revised version of *The Skin of Culture* for the Chinese readers, de Kerckhove (2020) writes,

> Satellites ignore national boundaries and replace our customary land-based psychology with one that is predicated on large techno-cultural fields. . . . The world affects me directly, but I too can affect the world. Usually people talk about a "good" or a "bad" attitude as an epiphenomenon, as intonation is to speech. But it should be the other way around: attitude is the message. It is one of the most

immediate and readily available means to manage the interval that is opened between self and world.

The symmetry between a (technological) holistic point of view and an immanent (human–environment) *point-of-being*[11] (de Kerckhove and de Almeida 2014) is what gives substance to the idea of a new sociotechnical paradigm in the making. Here, it is also worth recalling Pedro Alejandro Basualdo's (2011) words on "individual global responsibility":

> Individual attitudes and actions constitute a new form of global action that seeks decent answers to human suffering. Future generations will judge us by how we have reacted to a crisis that has challenged our scientific community and evidenced the scope and limits of our ethical and moral world. It is time to grasp the value of humanity as global citizens and awaken our individual global responsibility.

Today, Basualdo's call for an "individual global responsibility" has to keep at its core the critical questioning of what the "individual" is (not least because of the rise of AI): Is it human(s)? Technology(ies)? Environment(s)? A systemic merging? By delving into the coming to being of the quantum ecology, the goal of this book is to rework de Kerckhove's notion of attitude and inscribe it within an ecological framework, which is also a prompt to global action. Our aspiration is to bring about change in the world, but first of all, we hope to bring such change in the way people think of themselves as a whole and as actors of the world-sensing of which they are part. Today, we need this change more than ever, if not for us, at least for future generations.

1 Language: Alphabetics and Logographic World-Sensing

Chinese tend to think of language as something being written; our language is the writing. Language to a Westerner is spoken.
—Kenneth J. Hsü (1999)

Whether you read letters or ideograms, you will experience the world differently. And your relationship with language will be different too. Script forms determine whether the text represents speech or concepts. In its alphabetic form, the text you are reading now represents speech and does not require that you bring up the context. It is self-sufficient and, quite literally, contains the message. All you need to know are your language and your letters, and reading the text will then evoke the context. At the basic level of deciphering, the alphabetic sequence is also detached from the content of the text. You could say that it is "content agnostic." But if the text shows pictures, you simply cannot decipher the meaning without bringing the context forward. Like pictures or drawings, ideographies are *translinguistic*. If you are obliged to bring the context merely to be able to read the text, that fact translates into a very different world-sensing. All writing systems, even syllabaries and consonantal scripts, especially graphic-based ones such as Chinese, bind the readers simultaneously to the text and the context of the signs even before they can get to their meanings. And this is where East and West part company. The main difference is that phonological literacies detach the reader from the text. This specific relationship, repeated every time one reads, establishes a critical distance between the reader and the text. The text gradually becomes perceived as "objective" (not necessarily its meaning but merely its presence "in the world") and the reader's interpretation as "subjective."

Thus, a critical distance is extended to the relationship between the person and the world, also isolating the reader from the community. Because iconic texts involve the context first, that separation is not necessary. On the contrary, the reader is bound to the text and, consequently, bound to the context, that is, the community and the world. We intend to present that argument to document and illustrate our ground thesis of this chapter, namely, that writing media determine different personal, social, and political experiences and behaviors. They function as *operating systems* of culture. And when they change, all the applications, that is, the institutions that depend on them, change too.

On Language as *Dispositif*

How has language become the privileged means of communication of humans for thousands of years? How has it emerged as an organic control *dispositif* charged with meaning? Natural language is a characterizing feature of humans. It can be defined as a *dispositif* to the extent to which it constitutes a complex systematization of the human potential for vocal and gestural communication. Such a systematization can be seen as an embodied (organic) medium used to better cope with the environment and as a way for conveying human experience so that it can be shared, agreed upon, or contested. On this, Maturana and Varela (1987, 211) have written interesting pages, not only subverting the traditional relation and distinction between observer and observed but also unveiling the situatedness of any linguistic act:

> Observing arises with language as a co-ontogeny. . . . With language arises also the observer as a languaging entity; by operating in language with other observers, this entity generates the self and its circumstances. . . . And meaning becomes part of our domain of conservation and adaptation.

As a device of communication, language is an entangled affair of observation that brings forth ways of doing and being in the world. As soon as one acknowledges the (organic) embodiment of language, then to speak of its "naturalness" is somewhat misleading as this standpoint risks erasing the materiality of language. In fact, language brings with itself varying degrees of formalization and abstraction resulting from long-term ongoing adaptation among humans as well as between them and their surroundings. Fritjof Capra (1975, 33) notes, in this respect, that "ordinary language is a map which has a certain flexibility so that it can follow the curved shape of the territory to some degree." Capra then compares language to mathematics, arguing that both construct a formalization of the world, and yet, while the former maintains a connection—stronger in the case of ideograms, weaker in the case of alphabets—between the system of signs and the communicated experience, math's adoption of a more rigorous and necessarily abstracted systematization of the world leads to "a point where the links with reality are so tenuous that the relation of the symbols to our experience is no longer evident. This is why we have to supplement our mathematical models and theories with verbal interpretations" (33).

Capra's acute parallelism shows the inevitable negotiations that language-based and math-based systematizations of reality demand. On the one hand, the process of translation between math and language is problematic; it has always been. After all, every translation is always an adaptation and betrayal of the "original" (purposely in quotation marks because it is a contested notion): the root of the Latin word *traducere*, which literally means to "bring across" and from which the Italian *traduzione*, that is, "translation," comes from, is the same as *tradere*, which means "to deliver/ to give," from which Italian *tradimento*, that is, "betrayal," derives. On the other hand, Capra seems to suggest that language—as the ever-changing result of biological evolution as much as cultural contextualization—can be regarded as the most refined means humans have to map "reality." But certainly not the only one or the "best"; at stake is rather an issue of *fitness*.

Beyond these differences, the roots of both these *dispositifs* are traditionally in the phenomenological world: while this might be intuitive for languages, it also applies to math. Paolo Zellini (2022, 285) in this regard writes that "in math there are axioms, theorems, functions, and algorithms. But we might even refer to images or visions preceding the axioms and analytical formulas. In a process of progressive rarefaction, the purity of the sign or the mathematical image stands as a *limen* between visible and invisible" outlining an abstract landscape that has the strength of demonstrable laws (cognitive-analytical value) but also a loss of sense and intuitive comprehensibility (material-referential limitation). It was mathematician Jacques Hadamard (1949), in his *Essay on the Psychology of Invention in the Mathematical Field*, who discussed how mathematical thinking is visual in nature, rather than linguistic, creating an epistemological hiatus between these two *dispositifs*.[1]

The point is: both language and math produce two alternative configurations, based on different functioning logics and with different results (and purposes). Language leans toward the metaphorical while math leans toward the symbolic, but both metaphors and symbols are sides of the same coin: a way of "fictionalizing" reality in the etymological sense of the verb, that is, "to model" (Zellini 2022, 132). This also implies that we cannot content ourselves with remaining within the realm of the representational (cf. Mitchell 1994): what we claim is that each *dispositif* consolidates the internal consistency (not necessarily cohesiveness) of its own world-sensing, thus having effects that are very much performative. It is these effects that demand attention, aware of the fact that each world-sensing is always open and incomplete, even when a wide consensus can be reached.

The Inadequacy of Language and Mathematical Images

When the hiatus between language and math enters the quantum realm, their incommensurability becomes evident in that these *dispositifs* can only meet on a speculative ground, which does not share a common phenomenological referent. On this point, Burwell (2018, 51) writes that "a reliance on nominal language containing conceptual metaphors founded on the notion of substance and entity renders misleading any attempt to describe what amounts to potentialities of being, rather than the state of being itself" in that "in quantum physics elementary particles 'exist' only as a set of relations constituted by change, process, and events." Language already (and inevitably) provides a "chopping up" of "reality," and it does so through a meaning-making process that is metaphorical, not ontological. On the other hand, math symbolic formalization, while providing a model of physical reality that is testable and replicable, is not necessarily truer than other (linguistic or figurative) configurations—although certainly *fitter* to express behaviors that elude phenomenological observation. Were math to remain fully isolated, it would be simply pointless, in the same way as hieroglyphs remained for centuries "obscure" until they were deciphered via a process of translation. In fact, the possibility of a meaning-making and meta-referential *dispositif* beyond language is in question. Paolo Fabbri (2001, 21) contends that "it is possible to think that, independently from a strictly linguistic expression, a certain cognitive organization is in the making." On this point, Hofstadter (1979, 174) has written acute reflections, identifying different layers needed for deciphering out-of-context messages: (1) the frame message, (2) the outer message, and (3) the inner message. This nestled approach to communication questions the very link between signs and reality, or better the process of signification of the world. Physicist Bohm (2002, 27) notes that

> if we supposed that theories gave true knowledge, corresponding to "reality as it is," then we would have to conclude that Newtonian theory was true until around 1900, after which it suddenly became false, while relativity and quantum theory suddenly became the truth. Such an absurd conclusion does not arise, however, if we say that all theories are insights, which are neither true nor false but, rather, clear in certain domains, and unclear when extended beyond these domains.

It is not surprising that similar epistemological concerns also connoted the writings of the founding fathers of quantum mechanics, when the new discoveries threatened to leave the whole discipline conceptually groundless. Bohr (1949), for instance, noted that "no content can be grasped without a formal frame and . . . any form, however useful it has hitherto proved, may be found to be too narrow to comprehend new experience," leading to the need for "widening the frame." Hence, while maintaining rigor, hermeneutical flexibility is welcome, and distortion at the intersection of language and math becomes inevitable, even a productive necessity. Proof is that, even today, there is no unanimous agreement on what the math of quantum mechanics *means*, that is, what it stands for, and this is so because linguistic descriptions of quantum phenomena can only provide proxies to make sense of observations for which neither language nor human classic understanding of reality are equipped for. As soon as a *dispositif* becomes complex enough to allow for meta-referentiality, it also "digs its own hole" (Hofstadter 1979, 464), and in order to have a full account of it, it is necessary to step out and look at the inside under a different frame.[2]

Burwell (2018, 50) writes in this respect that, although having different positions, "Schrödinger, Bohr, and Heisenberg spent so much time talking and writing about what quantum physics revealed about the nature of language, what language revealed about quantum physics, and why any attempt to describe quantum concepts in language was likely to end in failure." Specifically, Heisenberg "approached the language problem from a historicized perspective; he believed that while the axioms of mathematics expressed an enduring truth, the limitations of (classical) language might be viewed as temporary" (55). Heisenberg then proceeded to call for a new linguistic toolbox able to account for quantum counterintuitiveness. However provocative his call was, from a semiotic point of view, this was flawed in two respects: language is a fundamentally phenomenological medium, so if "words do not fit," this can hardly be a temporary problem; further- more, Heisenberg failed to recognize that the historicity he attached to lan- guage applies to math as well. If math seems to express enduring truths, it is simply because, as a highly abstracted symbolic system, it appears as disembodied and universal. Schrödinger (1957, 71), instead, made of such embodiment a key point, observing that "[quantum mechanics] funda- mentally disrupts the manner in which spatial orientation is fundamental to the metaphors that we create, connected as they are to the sort of bodies that we have and how they function in our physical environment." In fact, the way to think of, elaborate, and apply math reveals that math too is a socially constructed practice, both symbolically and cognitively. On his part, Bohr acknowledged that the very idea of "reality" is just a label that is adopted for the sake of understanding: "We are suspended in language in such a way that we cannot say what is up and what is down. The word 'real- ity' is also a word, a word which we must learn to use correctly" (quoted in Petersen 1985, 32). In other words, "reality" is being created through communication, which can be seen as an act of (linguistic or mathemati- cal) measurement: after all, the Latin *definire*, which means "to define/to name," means exactly to trace boundaries, to mark where something ends (*finire*), and thus also, by extension, to decide what something is (or is not) establishing the codependency between observer and observed.

This does not mean to fall into pure arbitrariness but rather to acknowl- edge that "the laws behind the number, whose origins remain enigmatic, shall be inscribed in a much wider context, which includes any linguistic act, any possible sign or forms of expression, of which is clear the deficiency and inadequacy before the infinite sea of possible meanings" (Zellini 2022, 332). In fact, research suggests that the origins of numbers are not to be found into a transcendental realm but as a result of a cognitive-ontogenetic faculty (Dehaene 2011) at the basis of the differentiation between self and the world (not unique to humans).

"Reality" is just a word implying different ongoing fabrications depend- ing on certain bio-sociotechnical conditions. Hence, terms like *observation* and *apprehension* do not point to self-contained acts but rather are approxi- mations of knowing, a practice summoning its own embodied space-time horizon of existence at the very moment it is being performed. In this sense, knowing does not necessarily depend on human agency; it can be enacted by any information processing agency, organic or mechanic. Comparing machines to humans, Hofstadter (1979, 471) builds a significant parallelism in this regard: "No matter how a program twists and turns to get out of itself, it is still following the rules inherent in itself. It is no more possible for it to escape than it is for a human being to decide voluntarily not to obey the laws of physics. However, there is a lesser ambition which is possible to achieve, that is, one can certainly jump from a subsystem into a wider subsystem." Leaving aside whether this still applies to today's AIs, this passage helps to understand that, regardless of their metaphorical and/or symbolic configurations, all systems of signification are relative; all need to be translated across other systems so that their blind spots can be unveiled. The point to make here is that the evolution of language is that which has fostered a specifically meaning-making ecology, and yet, the world-sensing invoked by *dispositifs* other than language cannot be simply dismissed because they are unexplainable. They need to be embraced *qua* unexplainable.

At stake is the unpacking of the connections, overlaps, and differences between (abstract) *cognition, perception* (of the world), and (linguistic or numerical) *expression*. Studies in sociolinguistics (e.g., Alverson 1994; Lakoff and Johnson 1999), as well as on the history of mathematics (Serfati 2005), hint to the fact that, rather than a circular loop between these three realms, the bond between them is one of codependency. To speak of codependency does not mean to deny that these three realms are uniquely extant, but to question the link between being and (physical) existence. On the one hand, not only the physical world but also mental ideas are extant (albeit according to partially different onto-epistemological conditions), and both come into being always through a certain observing stance (be it human or not); in the same way, language and math are as much subjective as shared social systems of signification. On the other hand, these three realms mutually interpenetrate, and yet they are neither isomorphic nor coextensive (not to talk of altered states of cognition or dreams). One can think of something that is logically possible in abstract, and yet totally unplausible in the world, as well as of something that is not logical, yet concretely possible (especially when new scientific discoveries such as quantum physics arise), and discuss all these cases through different expressive forms, each one establishing its conditioned horizon of knowledge (cf. Priest 2007). Hence, to speak of codependency means to take a perspective *from within* that overcomes both analytical dichotomies (e.g., mental vs. physical) and privileged standpoints (e.g., objectivism) in favor of the exploration of the onto-epistemological conditions enabling and enabled by these three realms.

As well as claiming objectivity, the text of mathematics further detaches the reader from the context. But it does require a thorough mastery of the symbols and equations. Writing, in its turn, formalizes language and generates a number of applications that evolve into institutions.

Archaeology of Language (and the Body)

To retrace the origins and genesis of natural language is an endeavor that has interested scholars for centuries. In particular, since the second half of the nineteenth century, with the systematic emergence of paleontological studies and the initiation of neuropsychological research, this endeavor has led to the elaboration of various theories (not rarely in contradiction with each other).

Two major strands stand above many less convincing theories: one takes an anthropogenic perspective that sees language as the result of the evolutionary process of the human species, and the other favors an ontogenetic approach that places that evolution within the life cycle of the single human person. Debates among the supporters of the first option concern the major turning points along anthropological evolution, as well as the enabling conditions and the modus operandi of such process: for instance, Michael Tomasello (2008) stresses the dependence of language on gestures, while Quentin Atkinson (2011) points to an eminently phonological genesis. To bring them together, one could invoke (with due calibration) Giambattista Vico's (1744) hypothesis, which suggests humans first expressed their needs by grunts and cries and accompanying gestures and gradually, as human communities grew larger and more complex, felt the need to name things and express commands. Today, as Dietrich Stout and Thierry Chaminade (2012, 76) clarify, "Current evidence and interpretation supports and refines various 'technological hypotheses' positing neural and evolutionary connections between language and technological praxis." This not only stresses the inextricable intertwinement between the use of tools, neurological–physiological evolution, and the development of language functions but, more radically, supports the idea of the technological nature of language as an organic, embodied apparatus of signification.

The ontogenetic standpoint, instead, focuses on the innate emergence of linguistic faculties within the human body itself, that is, from birth throughout the first few years of life. Along this strand, controversies have to determine whether language can be considered a largely independent universal faculty or, by contrast, a chiefly cognitively and/or context-dependent process. For the latter, the two major exponents are, respectively, Jean Piaget (1936), who maintains a largely cognitivist standpoint, and Lev Vygotsky (1986), who defends a social-constructivist position. For the former, there is a line of thought that considers the origin of language an abrupt event in the evolution of the human species (Pinker 1994; Chomsky 1996) and one for which, by now, the human brain is innately predisposed, so that one can posit the existence of a universal grammar.

For the present discussion, an anthropogenic perspective is favored, allowing one to explore language as an enframing technology for providing a formalized mediation of/in the world. The work of French scholar Leroi-Gourhan (1993, 265), especially his study *Gesture and Speech*, is seminal in this regard in that it actualizes the discussion by bridging it to the contemporary situation:

> We already know, or will soon know, how to construct machines capable of remembering everything and of judging the most complex situations without error. What this means is that our cerebral cortex, however admirable, is inadequate, just as our hands and eyes are inadequate; that it can be supplemented by electronic analysis methods; and that the evolution of the human being must eventually follow a path other than the neuronic one if it is to continue.

Apart from envisioning very lucidly the biotechnological breakthrough humanity is witnessing today (the book was originally written in 1966), this passage is a good example of the historical-materialist approach of Leroi-Gourhan to the evolution of human faculties. As Gilles Deleuze and Felix Guattari (1987, 449) argue in *A Thousand Plateaus*, Leroi-Gourhan "has gone the farthest toward a technological vitalism taking biological evolution in general as the model for technical evolution." This means that humans and technology (and the environment in which they are immersed) are conceived as an ensemble. Witness the fact that the two volumes dedicated to this topic are respectively subtitled, in the French version, *L'homme et la matière* (*Man and Matter*) and *Milieu et technique* (*Place and Technique*). Indeed, what Leroi-Gourhan argues in *Gesture and Speech* is that the evolution of language is anthropogenically situated and context dependent, or, differently said, language, as the result of an evolutionary process, is an organic control *dispositif*. Nicolas Evans and Stephen Levinson (2009, 446) summarize the point arguing that "language is a bio-cultural hybrid. . . . Human success in colonizing virtually every ecological niche on the planet is due to adaptation through culture and technology, made possible by brains gradually evolved specifically to do that."

From an ecological point of view, to claim that technology is and has always been embodied means, on the one hand, to de-emphasize the distinction generally made between biological and nonbiological realms, or nature and culture. "What counts instead," as Erich Hörl (2017, 5) notes in the introduction to *General Ecology*, "is precisely to pass through the radical Nothing of technology, to question anew the relation between technics and sense, and to reassess this difference for the age of the technological condition." The advent of any new *dispositif* requires exploring how it is enframed by as much as enframing (for the very fact of being prosthetic) collective society and the individual, as well as how this double-sided actualization reverberates throughout and changes the horizon of reality. On the other hand, to acknowledge the technological nature of language and the codependency between humans and tools means also to relativize the "naturalness" of the human body. On this, Maturana and Varela (1987, 34) clarify that "all knowing is an action by the knower, that is, all knowing depends on the *structure* of the knower [emphasis added]." The idea of structured knowledge points to an entanglement between knower and known or, better, to the idea of knowing as an embodied process. In other words, the body itself can be regarded as a *dispositif*, in the sense of an apparatus—organic, for humans; artificial for machines—with its own affordances, which allows for the shaping of experience (and its meaning-making through language) in a *certain* way. Today, this approach takes the contours of a proper material engagement theory, of which Don Ihde and Lambros Malafouris (2019) are two main proponents: "we are *Homo faber*," they write, "not just because we make things but also because we are made by them." By subverting the subject–object relation, Ihde and Malafouris arrive to unveil the technologization of the "human," which will become increasingly evident with/through embodied technologies (Neuralink is just one case), as these will literally redefine what it means to be human.[3]

By opposing the universal or purely cognitivist ideas according to which it was the development of the brain that drove the evolution of the human species, Leroi-Gourhan sustains that evolution is a bottom-up process, so to speak, one that comes from the ground and moves up. For instance, according to Leroi-Gourhan, it was reaching the standing position that allowed hominids to develop increasingly refined tools, as two limbs were liberated from other motion duties. Similarly, the standing position, which triggered frontal bilateral sight, could also have "liberated" the mouth from the most impeding survival need, that is, to serve as the interfacing organ for the quest and ingestion of food. Hence, a space opened for language to develop as a representational-enactive *dispositif* to reach out, that is, to produce meaning and prompt action. Language, in other words, is a technology of the mouth as much as tools are technologies of the hands. Eventually, according to Leroi-Gourhan, the standing position and the need for eye–hand coordination brought about a physiological–functional modification of the skull and its volume (not vice versa, as often assumed).

Beyond the somewhat structuralist approach of Leroi-Gourhan, the link between gesture and language, which is at the base of an anthropogenic perspective of the evolution of language, is further reinforced by recent research showing that verbal language and sign language depend on similar and neighboring neural structures (MacSweeney et al. 2008). Moreover, in humans as well as in primates, gestures seem to have a direct effect on vocalization, suggesting that the refinement of vocal modulation all the way down to oral languages might have roots in sensorimotor activities (or at least be affected by them) (Aboitiz 2012), to the point of suggesting vocal-entangled gestures (Pouw and Fuchs 2022). In fact, the findings coming from studies on primate cultures have shown that *Homo habilis* was able to retain tools, which means not only to create and use them but also to circulate and transfer their embedded technological know-how to others—and be shaped by these artifacts. This has led to the suggestion that *Homo habilis* had already developed some forms of language for communicating with peers. In this view, language was used for communicating chiefly survival strategies, control, and command. How has this dimension evolved into the complex communication systems to this day? To find an answer, we have to look at the technologization of language, beginning with writing: we will do so by focusing first on alphabetic and then on logographic systems.

Archaeology of Writing

Walter Ong (1986, 35) famously claimed that "writing . . . is the most momentous of all human technological inventions." This is so because the fixation of sounds on supports that could overcome the time and space of oral contingency contributed to fundamentally restructure human cognition. The most interesting studies in this regard come from the scholars of the—informal—School of Toronto. Among others, Harold Innis (1972), Eric Havelock (1976), Marshall McLuhan (McLuhan and Logan 1977), and later Derrick de Kerckhove (de Kerckhove and Lumsden 1988; de Kerckhove 1997a), who continued McLuhan's work into the internet era, have all delved into the investigation of the origins of writing—not only alphabetic writing but also ideograms—starting from a literary–philological perspective. Their works have significantly contributed to the understanding of how different writing systems are responsible for (and, in turn, are shaped by) different and precise world-sensing and apprehensions of "reality." As Dan Sperber (1990, 42) succinctly put it, "Culture is the precipitate of cognition and communication in a human population."

Although symbolic representations (not only drawings) can be found on caves dating as far back as 45,000 years ago, these forms of fixation are still highly polysemantic and do not maintain any stable connection with a precoded linguistic system. Hence, to find the first prototypes of writing, one has to move considerably forward in human history. According to the work of paleontologist Denise Schmandt-Besserat (1977), forms of proto-writing can be linked to Sumerian trading activities and the dissemination of contractual practices (around 4,500 years ago) that involved inserting contractual clay tokens to represent merchandise (cattle, wood, wheat) in sealed clay bullae destined to be broken and thus to release the tokens once the contract had been fulfilled. Later the traders found it more convenient to imprint the shape of the tokens on fresh slabs of clay, creating signs that eventually evolved into cuneiforms (Logan 2004). The origins of Western writing are to be found, therefore, in the need to formalize (the method and the value of) daily trading practices.

From here began the slow process of abstraction that led the Phoenicians first to adopt the Sumerian syllabic cuneiforms and later, stressed by the pressure of their Egyptian taskmasters in the turquoise mines of Petra,[4] to develop a "secret code" of their own. Namely, Phoenicians elicited the twenty-two letters of their own writing system by borrowing the (acro) phonetic markers that the Egyptian scribes employed, much in the way Chinese ideograms use phonetic signs, to distinguish the meaning attributed to pictographic homonyms. Thus did the Phoenicians create the first alphabet, that is, a parsimonious set of signs that bore no longer a visual (iconic) relation with the intended object but a phonemic transcription of the word it designated. That said, it would take another refinement to achieve the Greek alphabet, namely, adding signs for vowels in the consonantal Phoenician script. The takeaway from this cursory survey is that the emergence of the alphabet, a momentous invention that is presently used by 70 percent of the world population, was in fact the result of an "accident" that brought together the traders of two very inventive cultures with complementary but incompatible linguistic structures, one where the lexicon was reserved to consonants and the other where vowels were also necessary to disambiguate the words.

Over the millenaries, different kinds of writing systems have emerged among different human settlements and were each conditioned by the specific typologies of the language they represented, monosyllabic where the syllable is generally also a full word (Chinese[5]) or polysyllabic (Indo-European, Hindi, Japanese, Korean), consonant based, that is, languages where the role of vowels is only grammatical (Semitic, namely, Egyptian, Phoenician, Hebrew, Arabic) or vocalic based, where lexical differentiation requires both consonants and vowels as in all European languages, including Basque, Hungarian, and Finnish. Each type required a different strategy to represent them.

As Yuxin Jia and Xuerui Jia (2005, 151) observe, language and writing are strictly interdependent: "The culture of a given society shapes its writing or orthographic system according to the features of its language. The writing system, in turn, influences the culture which creates it . . . as an active force, it promotes and reinforces its socio-cultural reality, the established mode of thought in particular." On this point, the hypothesis advanced here—both theoretical and methodological—is that writing and culture are interdependent to such an extent to be, in fact, *codependent.*

Whereas polysyllabic languages lent themselves quite naturally to phonological transcription, monosyllabic did not, on account of the extreme difficulty of disambiguating the great number of homonyms unmanageable by phonological rendering alone. The options that eventually led to a durable adoption of the most relevant strategy for each language depended upon how this evolved in context and over the centuries. As Janet Pierre-humbert and colleagues (2000, 12) note, "The vowel space—a continuous physical space rendered useful by the connection it establishes between articulation and perception—is also a physical resource. Cultures differ in the way they divide up and use this physical resource." This means that language and culture are an entangled matter not only at the level of con-ceptualization, or of meaning, but in the very fact of cutting through the physical *qua* perceptive. Bohm (2002, 82) explains this well: "Language [is] a particular form of order. This is to say, language not only calls attention to order, it is an order of sounds, words, structures of words, nuances of phrase and gesture, etc. . . . Every language form carries a kind of domi-nant or prevailing worldview, which tends to function in our thinking and in our perception whenever it is used." The material, the cognitive, and the perceptual are three configurations of the same whole: "All we can do," Maturana and Varela (1987, 242) write, "is to generate explanations, through language, that reveal the mechanism of bringing forth a world." To be entails a traversing of the physical: although from a complementary numerical perspective, Zellini (2022) reminds us that "cut" is the exact word mathematicians, since ancient times and in different cultures, have used to make sense of the discrete—of what "is"—as a splitting of the continuum.

In turn, each script form became the source of various applications, some of which bear critical importance for the main argument of this chapter. How languages were eventually written down has profoundly influenced psychological, social, economic, and political behavior. To give an example of this relationship, we aim to compare some of the key effects of alpha-betic literacy with those of Chinese logography, an endeavor that we hope will support our later examination of the effects of digitalization and, for admittedly different reasons, quantum technologies.

The Epistemology of Writing Systems
Language, as a meaning-making autopoietic *dispositif*, is simultaneously a credible source, if not a condition, of a way of thinking and of situating oneself and dealing with others in context—that is, a specific epistemology. Different theories (Devitt and Sterelny 1999; Barrett 2014; Lupyan and Dale 2016) have arisen regarding the relationship of language with epistemol-ogy, the least of which is not the controversial "Sapir–Whorf hypothesis"

(cf. Koerner 1992) that posits that language *affects* cognition. From this understanding, different languages—and the systems invented to visualize them—would develop different worldviews in the traditional sense of the term (we will see later that the question occupied and divided also ancient Chinese philosophy). This hypothesis generated as much adherence as resistance. Among famous naysayers is Steven Pinker (1994), who has gone as far as positing the existence of a universally shared language of thought he calls *mentalese* that, according to neurolinguistic research, would precede, not follow, wording. To which Michel Foucault could oppose his theory of *le discours*, with a focus on what he called "episteme," that is, a politically charged system of thought control that espouses various contexts and is subject to evolve into different formats over time and in different places.

The reason to bring up the Sapir–Whorf hypothesis is to introduce a significant difference between speaking and writing in their relationship to thinking. Writing externalizes, formalizes, and silences speech, establishing a separation and a distancing effect between the reader and the text (Havelock 1976). The question then is: how or why would written language influence the way of thinking more than the language it represented? Elaborating on the work of Stanislas Dehaene (2009) and Maryanne Wolf (2007), scholar Hye Pae (2020, 20) provides a working hypothesis:

> We can control our spoken language, but we cannot control the processing of written language at the moment of text exposure [reading]. Since the brain is rewired once reading skills are acquired through neuronal recycling, the extrapolation of the script→ brain restructure→ cognitive change is deemed reasonable.

This means that the acquisition of reading competences and especially writing skills dictates the coming to being of a precise mode of processing "reality." While speech may condition thought, writing is responsible for shaping a specific world-sensing.

Brain Issues

In his bestseller *Neuronal Man: The Biology of the Mind*, Jean-Pierre Changeux (1997) made public his research on synaptic epigenesis and how interaction with the physical, social, and cultural environment led to a selective stabilization of synapses in the brain that he called "mental objects." This research allowed his team to locate specific areas of the brain that stored three kinds of objects: *percepts*, *icons*, and *concepts*. A percept is the activation of a network of neurons by an external stimulation; the icon is produced by memory and arises from a selective stabilization of synaptic coupling between neurons; the concept is a skeletal image, a shortcut that refers to a similar but simplified neural stabilization. All such mental objects, also called "neuronal graphs," are subject to both individual and cultural predispositions. According to Changeux, percepts are guided by the senses and thus more vivid because they depend on direct sensory evidence, whereas icons being evoked by memory rather than immediate contact with the object under consideration have less precision or definition and thus evoke a more schematic image. Suffice to try it right now, looking at any object, then closing your eyes to summon it in your mind, instantly remembering what you were seeing. You will find that what you only remember is rarely, if ever, as vivid as the real-time visual experience. In concepts, a new level of abstraction would be reached separating the mental object from both sensory and iconic inputs. This approach may be useful to understand a core difference between reading phonological and logographic scripts. At the very least, it would suggest that reading phonological scripts would principally evoke concepts, while logographs would lean closer to icons.

This has also to do with the lateralization of brain functions. Increasing evidence (among recent studies, cf. MacNeilage 2008; Vigneau et al. 2011) has shown that the right and left hemispheres of the brain process stimuli differently: while the left side is better suited for rapid, small-scale, analytical processing, the right side favors large-scale, longer-duration, holistic processing. It is not surprising that with regard to language processing, evidence (Bookheimer 2002) has emerged for the left hemisphere playing a role in lexico-semantic tasks, while the right hemisphere is called upon for affective prosody and context-dependent meaning. Taking the cue from that, Pae (2020, 144) provides a neat summary concerning the neurolinguistic differences between logographic and alphabetic scripts:

> On a general scale, brain imaging studies show that written words are processed in different scripts, such as English, Chinese, and Japanese, through similar brain networks in the left occipito-temporal visual word form area, despite differences in the surface form of various written languages. . . . On a narrower scale, differences in brain specialization have been found according to scripts being read, despite the overlap found toward the left hemisphere at the global level. Chinese characters or Japanese Kanji tend to evoke greater activation and specialization in the left middle temporal region which is related to the mental lexicon, while alphabetic reading is likely to recruit activation in the left superior temporal

region and the angular gyrus which are related to the auditory processes through letter-sound conversion route.

According to Chinese linguist Yungxu Wang (2013, 247), the Chinese language presents higher usage efficiency (compared to alphabetic languages), but it also has lower acquisition efficiency (thus demanding longer time to learn it, whose process is eventually stored in long-term memory, LTM). This means that alphabetic languages are easier to learn but require more cognitive effort (short-term memory, STM) to use. As a consequence,

> The need for a powerful STM and higher inference capability driven by everyday language processing complexity has enabled the alphabetic language users to possess a larger and dynamic inference space in STM for complex problem-solving and creative activities. . . . On the other hand, the advantage of the ideographic language users with a larger LTM is helpful for cumulative and stable knowledge acquisition as well as systematic problem solving.

A closer look is needed to see how such differences could imprint the respective technocultural fields being informed by alphabetic or logographic systems. For instance, in their edited volume *Space and Time in Languages and Cultures*, Luna Filipović and Kasia Jaszczolt (2012) contend that, despite the (supposed) cognitive unity of mankind, the linguistic variations in the conceptualization and representation of space and time are immense, and they depend on highly complex (and embodied) processing. Such differences concern the way languages in general and their corresponding script forms in particular condition their respective cultures for the very basic representations of space and time, to which we intend to add the experience of selfhood to distinguish key features of Eastern and Western cognitive and emotional experiences.

Alphabetic Literacy

It is known that the Phoenician alphabet the Greeks adopted toward the end of the ninth century BC was written from right to left. Following suit, the Greek scribes, from epigraphic evidence (Guarducci 2005), began by writing to the left for fewer than one hundred years (c. 800 BC to c. 725 BC). But, after experimenting with writing alternately in both directions (a style called *boustrophedon* that was written from top to bottom in horizontal lines alternately and consistently reversing the orientation of each asymmetrical letter), toward the middle of the sixth century BC, scribes and stone carvers opted for the horizontal rightward direction, interrupting the line of script at the end of the available space as is customary today in print as it is in longhand. The name for that style was *stoichedon*, meaning "in the shape of (military) ranks" (cf. Austin 1938).

In a recent article, which responds to Iain McGilchrist's (2009) work *The Master and His Emissary*, in which McGilchrist deals with the hemispheric differentiation of the brain, Susanna Rizzo and Greg Melleuish (2021, 26) claim that, while the author provides evidence for a wide range of such differentiation, "[he] fails to explain *how* one hemisphere of the human brain, namely the left hemisphere, characterized by an analytical capacity and an orientation towards the factual, came gradually to dominate [emphasis added]."

Apart from the obvious fact that writing affects the visuospatial strategies of the brain, there is precious little on what exactly did provoke the relateralization of the reading/writing orientation. This is precisely the research question that de Kerckhove and Charles Lumsden (1988) addressed in their edited work *The Alphabet and the Brain*. The hypothesis (visualized in figure 1.1) is that if the ancient Greeks—and subsequently all Westerners—began to write in the opposite direction, it was because, by introducing signs for vowels in the Phoenician script, they perfected the fully phonemic representation of the spoken word, making the sequence of letters continuous as opposed to the batch form presented in their model. To decipher the new format of the sequence, the fully vocalic alphabet had the effect of soliciting from the brain an *analytic* strategy (emphasizing the left side of the brain, which maps

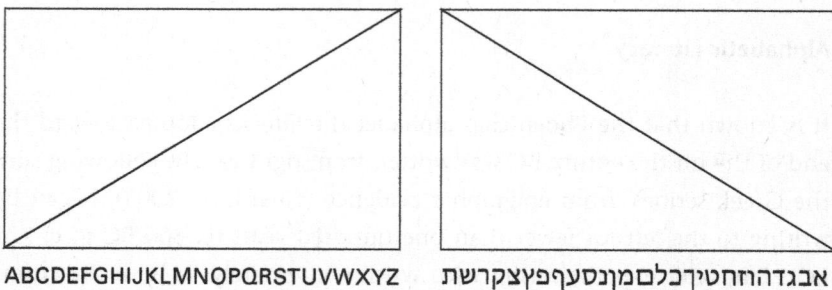

ABCDEFGHIJKLMNOPQRSTUVWXYZ   אבגדהוזחטיכלמנסעפפצקרשת

Figure 1.1
Whether your principal writing system is read to the right or to the left, you will tend to interpret differently the diagonals as rising or descending.

onto the right visual field via the optic chiasm) different from the consonantal scripts requiring *contextual* overview and guesswork (favoring the right side of the brain mapping onto the left visual field).[6]

This difference allows one to suggest that, from the adoption of the Greek alphabet onward, analysis, sequentiality, and many other operating features associated with the left hemisphere took the leadership in the collaboration of the two brain hemispheres and projected that strategy onto other cognitive and epistemological functions. The notion of *operating systems* proposed earlier helps to suggest how such changes happened over time. Fully phonological literacy established itself as the structuring process of social organization and instituted formal practices in law, governance, knowledge, art and architecture, and the military, as well as influencing changes in modes of thought and experiences of basic living conditions. One could compare the changes or innovations in different sectors of human activities to *applications* in networked computer environments. In other words, according to this hypothesis, the analytic strategy adopted to read alphabets influenced—although certainly did not determine alone—cognitive configurations in conceiving time, space, and also the sense of selfhood as applications of that specific operating system, not to mention the introduction of new structural and rationalist criteria for causal relationships in science and history, theoretical deduction in physics for atomism, modeling practices in geometry, anatomical analysis for medicine, and semiotic approaches to "reality" in philosophy. We shall now examine some of the main applications of alphabetic literacy.

Space

To appreciate space as a concept is not a given. It requires at least a measure of distancing between observer and observed, an effect of Greek literacy well documented by Havelock's (1976, 1982) seminal research. The suggestion that ancient Greeks experienced a new, if not the first, concept of space as a mental category can be supported by the development of theatrical space and the beginning of the appreciation of perspective, as well the building or rebuilding of cities in linear and parallel lines instead of the traditional concentric patterns. But the change is also reflected by new terminologies in the vocabulary of philosophers. Long before Isaac Newton, in his *Philosophiæ Naturalis Principia Mathematica* (1687), claimed that "absolute space, in its own nature, without regard to anything external, remains always similar and immovable," ancient Greeks such as Thales, Anaximander, Heraclitus, Democritus, Herodotus, Zeno, and a cohort of lesser stars had already elaborated a related concept, roughly three hundred years after the adoption of the alphabet and spanning over a period of 250 years more or less (between the sixth and the third centuries BC). By order of appearance, Thales of Miletus (c. 624–620 BC to c. 548–545 BC) was the first known philosopher and scientist of nature, an engineer by trade, who taught that nature was governed by fixed laws and not, as his contemporaries still believed, by supernatural forces. Anaximander (c. 610–546 BC) joined Thales's Milesian school and developed the notion of the *apeiron*, that is, "without limits"—in other words, the idea that space was infinite.

To Westerners, this notion, although not usually questioned by most people, is so intuitive as to appear irrefutable. Notwithstanding Einstein's general relativity, an example of a radically different experience of space would open this intuition to doubt. For example, Guy Deutscher (2010) in *Through the Language Glass* observes that in the Guugu Yimithirr language spoken by Aboriginal Australians, the perception of spatial relationships is "geocentric" by comparison to "egocentric" references in the West. This means that situating objects in space is related not to the position of the speaker but to their position in terms of cardinal points and relationships between them.

Thus, a rational, as opposed to organic, management of space in ancient Greece and colonies appears both in urban planning and in the formal architecture of theaters over approximately the same period, that is, from the end of the sixth to the fifth century BC. The life span of the most famous architect of antiquity, Hippodamus of Miletus (498–408 BC), covers more or less the same period as those of the main tragedians of the period (Aeschylus c. 525–c. 456 BC; Sophocles 496–406 BC; Euripides c. 484–406 BC). The cognitive effect of the theater was manifold. It first established a formal separation between the spectators and the spectacle, so much so that, except in comedy, for a tragic actor to address the audience directly amounted to breaking the rule of the convention. Furthermore, the theater also served as an always available physical image of consciousness, that is, a locus where meaningful symbolic action could take place. Frances Yates (1966) in *The Art of Memory* mentions that lyric poet Simonides of Keos recommended associating features and furnishings of the stage and entrances of actors in the play as a metaphor to teach his students how to memorize parts of their recitation. Here one could suppose that the internalization of conscious processes owed much to attending plays. Different from being a participant in a dance ceremony, as was the case for attending performances at the *odeon* where no formal seating was offered, the standard architecture of the Greek and later Roman theater actually educated the audience to spectatorship.

In this and many other developments of the alphabetic technocultural field, Romans and Athenians were thrust into *la société du spectacle* long before Guy Debord (1967) came up with the idea. Not that other cultures would not be put in the same situation by any ritual or any ceremony, but the difference is precisely in the level of participation required or not from the attendee. From Aeschylus to Brecht, the audience was asked to judge the play critically, *as if* in front of a written page. The invention of geometry and geography developed over the same period bears witness to a critical attitude to space that will later evoke different psychological intuitions from Simonides to Shakespeare's Hamlet (all the world's a stage . . .).

Time

Throughout Western history, the distinction between time and space has been a significant aspect of human cognition, in both public and private contexts. Early Greek philosophy established a clear separation between these two concepts, a distinction that has since been reflected in duration categories, historical records, and the philosophy of history for time, as well as geography and geometry for space. Benefiting from the distancing effect of literacy, time was formalized in identifying cause-and-effect relationships, on the one hand, and by segmentation and measurements, on the other.

Both addressing the passage of time, Heraclitus (c. 540–c. 480 BC) and Parmenides (c. 515–c. 485 BC) are frequently opposed on that issue, asserting, the first, that everything changes and, the second, that nothing changes—views from which one could conceivably derive simultaneously Newtonian notions that time is irreversible (Heraclitus) and the principle of conservation (Parmenides) that eventually would be expressed as the laws of thermodynamics. As Newton (1687) put it, "Absolute, true, and mathematical time, of itself and by its own nature, flows uniformly, without regard to anything external."

Since early Greek philosophy, time was clearly distinguished from space, both in public historical records, and in private psychological constructs, as in reflecting sequentially over one's own past (e.g., the autobiography of Xenophon). In personal reflections, the format of temporal sequencing was applied routinely to account for personal as well as historical life situations. Among the evidence collected about how ancient Greeks conceived of time past is the word *hupomnemata*, meaning "things to remember" and, by extension, "things to keep in mind." Michel Foucault (1997, 210) devoted careful attention to the practice:

> Hupomnemata, in the technical sense, could be account books, public registers, or individual notebooks serving as memory aids. They constituted a material record of things read, heard, or thought, thus offering them up as a kind of accumulated treasure for subsequent rereading and meditation . . . for a purpose that is nothing less than the shaping of the self.

Differently from confessional diaries, which contain the unspeakable, *hupomnemata* had the function to build an apparatus around and in support of the self by keeping track of the already said.

The evolution of Greek mythology is also instructive about how the idea of individual destiny was meted out to humans. The word for "destiny" was *ananke*, meaning "necessity." It was personified by the goddess Moira or also Aisa, and in later developments of the myth, Aisa has three daughters: Clotho, the goddess weaving the thread of destiny; Lachesis, measuring it; and Atropos, cutting it (cf. Kerenyi 1951). This apparently straightforward myth, by implied assumptions, contains a lot more than a parable of individual destiny. It assumes that time can be estimated independently from space; is irreversible (with a beginning, a duration, and an end); is unidirectional, going from left to right (in all sculptural or painterly representations, Clotho is on the left and Atropos at the end, on the right); and is measured not only in terms of how much is allotted to anyone but also implying that measuring it is a standard practice. It is, in other words, a very modern representation and conception of time. In chapter 3, we will see that ancient Greeks also maintained a qualitative conception of time—*kairos*—which today's Western cultures have lost and which is somewhat enmeshed with space, intended as the right time for an event to occur or an action to unfold.

Self

Among the main crucial effects of alphabetic literacy is the onto-epistemological condition of the private mind. Rizzo and Melleuish (2021, 38) support this idea by arguing that "there appears to be an undeniable correlation between the appearance and adoption of writing and the rise of analytical thought and self-consciousness." All writing systems, including logography, allow individuals to take possession of their content and store it in memory for their own use. The alphabet, however, confers that opportunity to a much higher degree because the simple fact of reading it has the effect of internalizing it, not as ideas or images, but as the sound along with the meaning of the language it represents. Thinking in phonological terms is a very different operation than doing so in terms of ideas or images. It offers many opportunities for verbal innovation by breaking down the complexity of imagining into modular and manageable units of words and concepts. Initially, alphabetic literacy was only a support for oral delivery and for remembering what to say—at worst, a corruption of *memoria* and *oratio*. In ancient Greece, as in Egypt, writing was the task of scribes at the service of powerful people who themselves could not read. All reading then was meant to be done aloud, underlying the fact that the alphabet was at first a sort of mnemo-technique to support the continued dominance of oral speech. Perhaps even at the very beginning in the mind of scribes, but surely with the diffusion of mechanically reproducible texts as printed codex, people began to read silently for themselves without feeling the need or the default condition to share the meaning of what they read with other people. Silent reading made private what used to be a public activity. Thus, reading silently also turned speaking into highly specialized thinking. Rizzo and Melleuish (2021, 31) observe that "once poetry and the similes it contained were congealed in the written text, they became the object of reflection and commentary. Through this process, human beings finally became aware of their individual existence and conscience, thus engendering a hiatus between the self and the world of experience."

What Rizzo and Melleuish call a "hiatus" is something much more important—to distinguish literate from logographic epistemologies—that Havelock (1982) called "the separation between the knower and the known" in his *Preface to Plato*. Here, he describes one of the key features of both the scientific process and the establishment of the common ground of judgment in Western cultures, that is, the clear distinction—and separation—between subjectivity and objectivity. Havelock (1982, 201) addresses the origin of the concept and the experience of the self, one of the most earnest preoccupations of Western philosophy and critical theory from Heidegger to Foucault and Derrida:

> A correct way of stating the effect of the revolution, if we are to employ modern terms, as we must, would be to say that it now became possible to identify the "subject" in relation to that "object" which the "subject" knows. Here we concentrate

on the new possibility of realizing that in all situations there is a "subject," a "me," whose separate identity is the first premise to be accepted before we pass on to any further statements or conclusions about what the situation is.

The rise of individualism in the West and the formation of the private self can be attributed to the effects of the alphabet that shifts the priority of meaning creation from context to text. Havelock continues, "The concept of the autonomous personality was not one that could be achieved in the abstract as though it were a scientific solution to a problem in external nature. . . . It was a personal and intimate discovery" (215). That "personal and intimate discovery" is what is called today the "private self" or, more generally, "privacy." We associate the formation of an internal, secluded, and protected ego with the appropriation of language, a communal medium, for personal use. Bringing together the arguments made above, this combination may indeed give rise to individualism as a psychological drive with far-reaching social and political consequences.

Interestingly, Havelock points out that the constitution of this private self is also "the source of all moral imperatives and all criteria of true and false" (215). Indeed, another direct correlation between literacy and the appearance of a private self was the emergence of the sense of guilt. The appearance of the private self changed the axis of responsibility from outer to inner, that is, from allegiance to the community to that of the person (i.e., to oneself). Greek tragedy could be a significant correlate because it shows guilt in action, so to speak, in all Aeschylus's plays (from *The Persians* to *Prometheus Bound*) first but even more precisely so in Sophocles (*Oedipus, Antigone*) and Euripides (*Medea*). From the stage at Epidaurus, fourteen thousand spectators were shown how to take responsibility for their own destiny instead of being exclusively committed to that of their community, such as it is exemplified in the earlier Homeric epics more attuned to "shame" cultures. Over the long run, interest shifted from emphasis on the community, represented by the tragic chorus, to focus on the individual actor(s) to the eventual exclusion of the chorus altogether in Euripides. Even the later stages of Greek mythology sustained the principle of individual destiny as separate from that of the immediate group, family, clan, and city. In fact, the gradual emancipation of the individual metamorphosing into a citizen is also a matter of privatization of its psychological attitudes toward the world, which goes hand in hand with the emergence of literacy. On this, Mikhail Bakhtin (1981) has written extensively making a case for the emergence of the written novel (and the codex book, as opposed to the epic, especially in its oral form) as a sign of the recalibration of the subjects' fate from the outside world governed by the supranatural to their inside psychological world (initially populated by phantasmagorical images), opening the way to the very possibility for individual change and growth.

Alongside with privacy that made it possible, guilt was profoundly encouraged by the sacrament of confession in Christianity as a practical method to impose self-control in a society where, at least among the literate ranks, everybody would acquire (at great cost of strife and bloodletting) the right to make up their own mind. Regarding Christianity, McLuhan occasionally reflected (albeit in conversation and not in print) on the possibility that it was itself an effect of literacy, along with Judaism and Islam, thus associating monotheism not only with "religions of the book" but also with a sublimation of selfhood in the idea of a personal god. Different from Yahveh, who relates to the Hebrews collectively as a people responsible for their actions (think of the need to find only eight "just" individuals to save Sodom and Gomorrah from Yahveh's wrath), Christ, issuing from the same culture but not adopted by the Hebrews, addresses the person's responsibility, individually. From its early beginning, Christian missions and evangelization bridged the gap between literate and oral communities, extending the personalizing effect of alphabetic literacy to vast numbers of people who could not read. Christian missions (Nestorian, Protestant, and Catholic) were also among the first sustained contacts between East and West with varying fortunes since the seventh century AD when it is said that the first translation of the Bible in Chinese was made.

Logography

To most people unfamiliar with it, the Chinese writing system appears impenetrable to all except to individuals born in the Chinese culture. Upon examination, however, it is quite intuitive and even logical for a language structure that attributes a specific signification to each syllable, a feature that has led some linguists to classify it as "monosyllabic" (and even "morphosyllabic"), even though this classification remains disputed. Tolerance for contraction in oral speech evolved over time, but without affecting the basic principles of the writing systems developed in each culture. Even though in Asian cultures there are various logographic systems (e.g., Chinese, Korean, Japanese), for the sake of the argument, we will concentrate on the evolution of Chinese logography, which is also the most adopted.

The Chinese Logographic System

The challenge of using single open syllables to designate words is that the number of open syllables pronounced by human speech is limited in Chinese to little more than four hundred distinct sounds (Pae 2020), but the number of words an advanced culture needs for daily use, not to mention the added complexities of abstract notions, scientific observations, and transcriptions of foreign words, numbers in the tens of thousands. To help disambiguate many homonyms, at least in spoken exchanges, four tonal values were added to the words that were spoken. In writing, only Pinyin (the official alphabetic transcription of Mandarin logograms) indicates the tonal value by placing diacritical signs above the vowels. Of course, tonal differentiation is not the only or even the principal means by which Chinese speech and writing disambiguate their homophones. Further to limit the number of ideograms, the most common method (90 percent of written Chinese words) is similar to the Egyptian determination practice of placing a phonological marker before the semantic icon.

The unique feature of written Chinese lies in the fact that the iconic radicals of logograms serve simultaneously as signifier and signified. This means that in Chinese, what Western linguists call the "double articulation"—the representation of meaning through both phonology (by itself meaningless apart from onomatopoeias) and morphology, the shapes that carry the meaning—is reduced to a single articulation, where only the morphological representation of meaning is present.

Hence, the examination of Chinese radicals helps identify the history of their concepts. On this point, Yuxin Jia and Xuerui Jia (2005, 154) further suggest, along the line of what we discussed earlier about brain functions subdivision, that "the historical development of Chinese characters was marked by the tendency towards a configuration in which the graphosemantic component became fixed on the left and phonetic component on the right. We assume that the gradual development and formation of this compound occurred in correlation with the division of labor in the brain." This approach is all the more relevant that over three thousand years since the evolution of pictograms, despite numerous opportunities, their shapes may have been stylized and their number reduced but the principle itself has not substantially changed. Chosen among the original 540 radicals of the archaic system, itself derived from earlier pictograms (c. 1200 BC during the Shang dynasty), the design remains more or less the same in the 214 graphs of the modern system, officially consolidated as the standard set of icons to determine meaning among homonyms.

Being constituted of ideograms and deriving from a pictorial system of signification, every Chinese character contains an explosive (holistic) force coalescing shape, sound, and meaning (cf. Taylor and Taylor 2014). While it is possible, in principle, to take them apart, their specification and function become pertinent only within the whole character. Shape, sound, and meaning, then, are not only interlaced but also consubstantial: as Jia and Jia (2005, 152) contend, "the graphemes, contours, and shapes of Chinese characters which themselves have content and messages were born out of the perceptive or visualized experiences." This is also why, in Chinese, calligraphy is so important and tones are made pertinent for signification and disambiguation. This applies to alphabetic languages only to little extent, the shape being technically irrelevant and the link between signifier and signified arbitrary. Moreover, the Chinese language is syntactically and semantically combinatory. For instance, there are no conjunctions and no morphological inflections or verb conjugations. On this point, Geoffrey Sampson (2015) contends that the Chinese is an isolating language: meaning, so to speak, is derived by accumulation, juxtaposing characters to what follows and precedes so that they only "make sense" in context and as a whole, synthetically. As an exercise, one might think that, while alphabetic notations embed morphosyntactic discreteness into writing continuity, Chinese logograms embed discursive continuity into visual discreteness. To this combinatorial feature is also associated the fact that, despite having a subject–verb–object construction, the meaning of a Chinese sentence can remain "suspended" until the end, where a particle tells, for instance, if the sentence is to be intended as a question or also if it is to be interpreted in the past. Of course, some alphabetic languages—especially subject–object–verb languages—do also bear a degree of uncertainty (as in German, at least from the point of view of generative grammar), but it is the mix of all these features together to eventually make the Chinese script prone to be cognitively processed differently from alphabetic systems. Notably, studies have shown that Chinese people tend to process information holistically

(Masuda and Nisbett 2001; Miyamoto et al. 2006) compared to people in the West, who tend to process information analytically and sequentially (White et al. 2008). In a similar vein, research has provided evidence that while Westerners tend to produce linear reasoning structures, Asians more willingly adopt circular rhetorical strategies (Kaplan 1966).

Logographic Effects on the Chinese World-Sensing

Perhaps, the radical and pervasive difference between the effects of alphabetic literacy and those of logography has to do with the reading process itself. Put simply, reading the alphabet detaches the reader from the text as well as from the context, whereas deciphering logograms requires getting fully immersed in context.[7] Hence, different characterizing epistemologies follow. In compiling the entry for *The Stanford Encyclopaedia of Philosophy*, Jana Rošker (2021) calls the Chinese epistemology "relational":

> The naturalistic epistemologies that prevailed in Western discourses were dealing with the external world (or objective reality), which was to a great extent independent from the subject of comprehension. Chinese approaches to knowledge can be called relational epistemologies, because they refer to relations . . . based upon viewing the world as a complex structure composed of relations, intersections and interacting feedback loops.

Cognitively, the consequence of this approach is that the act of knowing itself constitutes an intricate bond between the knower and the known, which transcends them both, by being inscribed into a holistic world-sensing.

Another corresponding aspect of Chinese epistemology is that cognition cannot be isolated or separated from emotion or moral responses (see "self" below). The organ of perception is not the mind as in the West but the "heart-mind," defined as *xìn*, which is where emotions, perception, intuition, and rational thinking cohabit. That said, in Homeric Greece, the source of cognition at first was called *thumos*, something relatively close to the concept of *xìn*, responsible for both perception and emotions. Progress in anatomical speculation, however, separated them by attributing cognition first to the liver (hence the punishment of Prometheus, see below), then to the lungs, and eventually to the head (*psuchè*), where cognition was isolated from emotion.

The question of the link between language and thought (and culture) was raised in China more than 2,500 years before Benjamin Whorf brought it to scholarly attention. Three main players, Confucius (551–479 BC), Mozi

(479–381 BC), and Laozi (likely sixth century BC) created three durable schools of thought. Confucius would be the determinist of the lot, attributing (perhaps for political reasons so as to ensure stability in the empire) a direct and binding relationship between names and functions of the state. Confucius could be said to have instated an entire "apparatus," *à la* Foucault, to establish strong institutional behavior bound to and by discourse. By contrast, the Mohist school considered that it was up to language to adapt to evolving social circumstances and not the other way round. The school of Laozi, who founded Taoism, recommended not to trust language's binding power altogether: "In Laozi's view," Rošker (2021) writes, "every linguistic concept is determined by time and space, and can therefore represent only a partial, incomplete expression of reality, which he saw as integral, dynamic and holistically structured."

Space

Influenced by Taoism (and Buddhism), Asian cultures foreground a balanced copresence of opposites and the symbiotic immanent inscription of humans into the world. According to sinologist Joseph Needham (1954, 458),

> The Chinese did not feel the need for [geometrical] forms of explanation—the component organisms in the universal organism followed their Tao each according to its own nature. . . . There is no more distinction between mind and body, subject and object. . . . We look around and perceive that . . . every object is related to every other object . . . not only spatially, but temporally.

This is a conception of space that is qualitative, that requires space to be lived through, rather than occupied. Space is always already natural, social, and political at once. On this point, Alberto Castelli (2015, 27) notes that the Chinese "did not compete with nature for they never had a future to decode. . . . Asians, China specifically, therefore did not challenge nature but rather accepted it." It is not that China remained oblivious of technological innovation: in fact, gunpowder and the compass, just to name two (beyond the press and paper), were known to the Chinese seven centuries before the Westerners. Yet, these innovations were inscribed into a different world-sensing, one that valued equilibrium over evolution and conquest.

Interestingly, the Chinese have also fostered the concept of *yǔ zhòu*, which defines the cosmos as a whole. Indeed, *yǔ zhòu* means more properly "space-time," leading to suggest that traditionally the Chinese, differently from Western cultures, have always thought of these two dimensions conjointly, in a way that, albeit unscientifically, gets close to that developed by modern physics. This vision is one that is inherently ecological to the extent it relativizes "being-in-the-world" seeking adaptation instead of imposing a mastering approach over nature. The philosophy of wholeness that the yin and yang doctrine posits favors the quest of harmony within the body (an example is acupuncture), as well as between the body and the surrounding. It is not surprising that China has long represented and named itself the "Middle Kingdom" (the characters that stand for "China" and are rendered in pinyin as *zhōng guó* mean literally that): this underlies not much a command vision but a balance-seeking one.

Time

It is often said that the opposite of linear time is cyclical time, a differentiation credited to the observation by many if not all early settlements that there was a regular return in the seasons and in the patterns and positions of the stars for the latter, while for the former, the intuitive perception of the continuous, uniform, and irreversible progression of time led to the deduction that it was an independent constituent of life and being. The Chinese conception of time is said to be related to the concept of the eternal return implicit in the yin–yang complementarity, as well as associated with a culture that remained predominantly agrarian for most of its history save the past 150 years. But it is more than that. Chinese patience is proverbial. From knowing how to wait for the "right" time—or *wúwéi*, the time of nonaction—rather than, as in the West, giving a kneejerk response to any situation, the Chinese experience another dimension of time that usually feels that "subjective" time is something to bear. Li Mengyu (2008, 67) explains that "Confucianism holds a flexible attitude towards time. It accentuates 'the right occasion' and 'the right opportunity' in dealing with affairs." Interestingly, when translated into language, this conception of time metamorphoses into a suspended apprehension of what is being referred to: as Castelli (2015, 26) puts it very simply, "Chinese language knows only the present, it does not have either past or future." This holds true also when looking at the Chinese conception of the universe, which, according to Castelli, is vertical and projected toward eternity by implicated repetition, so that, in language, the flowing of time remains largely semantically and morpho-syntactically unmarked.

This chrono-balance finds a reflection in the various ways in which Chinese people conceptualize and enact time through space or, better, how they give linguistic shape and embody time via spatial metaphors. At the intersection of thought and embodied experience, linguistic metaphors can be regarded as strategies to make sense of the world (Lakoff and Johnson 1999) and enact a *certain* "reality". This resonates with what is discussed above about concepts being a bridge between thought and physical reality, and yet, metaphors are particularly interesting in that they help to express abstract ideas in figurative ways and, in so doing, elicit an embodiment of these same ideas. As Burwell (2018, 40) writes,

> Conceptual Metaphor Theory proceeds from the argument that our everyday interaction with and observation of the world around us gives rise to image schemas—recurring, dynamic, multimodal patterns of perception that emerge naturally from our everyday experience—where "experience" is understood to have basic perceptual, motor, emotional, historical, social, and linguistic dimensions. Image schemas organize our perceptual experience, and thus give rise to understanding.

Most important, metaphors can be said to produce meaning while *maintaining* the superposition between thought and its concrete linguistic and gestural manifestation. Indeed, it is certainly true that the cultural context in which metaphors are adopted helps disambiguate how they are to be interpreted, but it is also true that literal and figurative meanings coexist, regardless of their disambiguation. Epistemologically, metaphors are complex affairs; they are Gordian knots of signification intertwining—and bringing such intertwinement to light—language, body, and thought.

For instance, studies (Alverson 1994; Ahrens and Huang 2002) have reported that, much more frequently than in English, in Chinese, speakers figuratively face the past, with the future behind their back. This impacts on the fostering of a more variable mental timeline, which is activated in context via gestural enactments (Ng et al. 2017). At the same time, it also happens that Chinese people represent time as being above (past) and below (future); this is usually prompted by congruent metaphors, thus proving that the latter do have an influence on thought and on the conceptualization of time, both in the short and long term (Lai and Boroditsky 2013). In other words, the way in which Chinese people give linguistic and gestural concretization of time via spatial triggers is isomorphic to their understanding of time and space as balanced and coextensive categories, certainly more balanced than the linear understanding manifested by speakers of alphabetic languages. Different language-based configurations of time have, in turn, inevitable implications on what humans consider causally related. If the link before–after is not an objective universal but a linguistic and embodied construct—as sociolinguistic evidence shows[8]—then the very possibility of establishing causal relations does not belong to facts but to how people "see" these facts.

Self

The "self" in traditional Chinese epistemology consists of three principal features: as seen, an organ of perception called *xin*, translated as the "heart–mind," including both cognition and emotion, as well as a "moral" component, or attitude that is tightly bound to the other two. In fact, this moral dimension is not merely added to the cognitive experience but integrated with it. There is no doubt that the presence and sense of self and self-awareness is just as strong and prevalent in China as it is in the West. But it is not alone. Not only does the Chinese person feel integrated within the environment—be it natural or industrial, the city or the country—but that integration is also shared with everybody else. This integration is constitutive of every cognitive engagement with places, times, and people.

This vision finds an emblematic realization in the teachings of Confucianism, which seeks stability and order through the replication of existing social and power relations. "Confucianism," Pae (2020, 121) notes, "emphasizes humanistic values in order to be in harmony with the law of the universe or heaven, including familial and social harmony, filial piety, benevolence, and ritual norms." To do so, "Confucianism favors past over present, orthodoxy and conformity over original and heterodox thinking, and the study of human relations and classic texts over the probing of nature" (Jiehong 2005, 109). While it is intuitive, based on these premises, to think of Confucianism as a conservative doctrine, there is more to that—namely, a collective-by-default understanding of the social realm (Hofstede 1980), based upon an integrated "I–We" relation. On this point, Richard Nisbett's (2003) historical observations differentiating ancient Greeks from ancient Chinese highlight that, while the former considered the utmost form of social life the attendance of theatrical plays (thus fostering a critical distancing between event and viewer), the Chinese favored family and friends' gatherings.

The homeostasis of the Chinese self within the context is linked to both linguistic constructs and visual representations. Concerning the former, linguistic analyses (Zhou 2001; Gong 2009) show that, in Chinese, time-moving metaphors (time moves and the subject is still) are more numerous than ego-moving metaphors (time is still and the subjects move along a given temporal axis): consequently, Chinese speakers are less likely to take an ego-moving perspective (than, say, English speakers are). Concerning visual representation, Castelli (2015, 28) acutely notes that "in the traditional painting landscapes and bucolic scenes precede people and if there are human figures they are very small compared to the economy of the picture." In a similar vein, Richard Strassberg (1994) discusses radical differences in the way in which premodern and modern Western and Chinese travel writers represent themselves and their journeys, with the latter group favoring, for instance, a bidimensional depiction of the experience in which the traveling "I" is, in fact, a mere glazing stance, while the former group tends to represent the journey as an ego-led and ego-moving experience. As Strassberg (1994, 22) writes, "The quest is less significant in Chinese travel writing," and therefore, "whereas Western travel writing, with its novelistic orientation, emphasizes social events and portraits of noteworthy characters, the poetic underpinnings of Chinese travel writing tend to stress objects and qualities perceived in the landscape" (31). The intertwining of cognitive and moral self likely hindered "objective" knowledge (knowledge not necessarily linked to morality), thus also preventing a positivist discourse and exploration of nature, in the sense of a mastering approach toward it. Nature has always appeared to Chinese people as a dimension to get along with, rather than one to dissect and tame.

Bridging the Gap between Technocultural Fields

It is at least since the Renaissance that the (Western) subject has started to claim an anthropological centrality within phenomenological reality (just think of Leonardo da Vinci's Vitruvian Man). It is exactly during the Renaissance, however, that this anthropocentric centrality coincided with a slow decentering process, effecting a gradual relativization of the subject's role in the big scheme of the cosmos (e.g., Nicolaus Copernicus's revolution). Not incidentally, it is at that time that a formal conceptualization of perspective was developed. Such conceptualization can be seen, on the one hand, as the psychological externalization of the subject's need to see oneself represented into the world, while, on the other hand, it unveils, by necessity, the relativism of one's own stance within the world. Let us note in passing that perspective also restated and confirmed what ancient Greek theater had partly accomplished before, that is, a formal separation between the object and the subject of the viewing experience, *theatron* meaning "that which is to be seen."

By comparison, perspective has not become a major driving principle in how the Chinese represent the world. Chinese painters of a similar epoch did know perspective, but the ethos that guided their apprehension of the world was different from that of Western peers. To represent the world, Chinese painting favors the wandering of the gaze in an all-encompassing vision. As scholar Kwo Da-Wei (1990, 70) contends, "The Chinese concept of perspective, unlike the scientific view of the West, is an idealistic or supra-realistic approach, so that one can depict more than can be seen with the naked eye." Chinese painters—we might provocatively suggest—were cubists *ante litteram*.

The formalization of perspective in the West was, at once, cause and effect of the emergence of a rigorous mode of thinking based on reason and calculation, which eventually disrupted the harmonic relation between humans and environment. Rationalism pushed Western thinkers, philosophers, and early scientists to observe and describe reality according to objective categories. This finds a clear example in René Descartes's positing of a distinction between *res cogitans* and *res extensa*: the former, which pertains to the thinking subject, is alive, animated, and ultimately able to know; the latter, in turn, is something that merely exists and is governed by deterministic laws (Newton's mechanical physics is indeed an example).

By contrast, as Li Jiehong (2005, 108) writes, "Science in China, from the very beginning, was congenitally deficient in rational thinking. . . . That's why the development of science in China lacked its own impetus and began to fall behind the West in the Song and Yuan Dynasty." Jiehong frames his position from the point of view of the history of science, arguing that it was "the popularity of dialectical thinking in classical Chinese philosophy" (108) to prevent open scientific interrogation and, for instance, the establishment of "the theory of the atom" (108), a statement to which scholar Kenneth J. Hsü (1999) adds that "the historical inevitability why Isaac

Newton was not a Chinese seems to have been rooted in an idiosyncrasy of a linguistic development." Physical reality, for ancient Chinese thinkers, was complete and self-sufficient, and the subject played an integral (fixed) role in it: there was no fraction or disruption as a precondition to scientific quest, only a harmonic state of things to be appreciated and defended.

On the other hand, without the application of the principle of analysis to the observation not of nature but of matter, the imagination of the atom could not have occurred anymore to "atomists" Leucippus (active between 450 and 430 BC) and Democritus of Abdera (c. 470–380 BC) than to contemporary Chinese scientists. A much-discussed fragment from Democritus of Abdera, quoted by Aristotle, Lucretius, and several others, recites, "Just as by combining the letters of the alphabet in different ways we may obtain comedies or tragedies, ridiculous stories or epic poems, so elementary atoms combine to produce the world in its endless variety." It seems that, by experiencing and applying the epistemology that comes with literacy, ancient Greeks adduced the existence of atoms without access to anything close to empirical observation. That said, comparisons and correlations are no proof that there is a causal relationship between the invention of the alphabet and the imagination of the "undividable"—the meaning of atom—but what is astonishing is the *modus cogitandi* linking the ancients to modern physics.

This did not happen systemically in China for the reasons discussed. Yet, in his article on multilinguistic alphabetic–logographic capabilities, Wang (2013) updates this scenario advancing the idea that, while "from cognitive linguistics and knowledge science points of view, the key properties of traditional Chinese might lack some powerful representation means to precisely and quantitatively express complicated systems and causalities" (255), nonetheless "bilinguals and multilinguals, particularly those with a combination of both categories of alphabetic and ideographic languages, may take significant advantage in simultaneous short-term memory and long-term memory development" (248). Most important, from our perspective, it remains to be seen the extent to which Chinese refractoriness to systemic and causal formalization might eventually prove advantageous—also through linguistic cross-fertilization—in the context of a new emerging quantum ecology, which inherently questions cause–effect links and determinacy.

There would be extraordinary value in bringing together the functionalities of both East and West operating systems. It would require not only that they agree to agree but also that they would learn each other's language, not merely trusting their automatic translating systems but ingraining language and writing systems in early schooling. The Chinese are already further ahead in that many more Chinese citizens speak English than Americans or Europeans speak Chinese, among whom many also do part or all of the higher studies in the West, whereas the Western student who attends Chinese universities is a much rarer bird.

The fact that Chinese researchers and technologists already benefit from using both operating systems is proven by the speed with which they select, adopt, and adapt Western innovations that serve them and how they rapidly improve on them: witness the internet and, among thousands of other borrowings, now their progress in quantum technologies. A recent example is that it is not China that invented the metaverse, but there are already over 1,500 Chinese companies that are using and implementing the concept, and many more are predicted to follow suit. Assuming that the West does not get the message to explore the alternative operating system, in no time would China overcome the West's alleged technological advance, but the real and long-term advantage of sustained collaboration would be lost.

A similar position to Wang's in support to plurilingualism comes from Bruno della Chiesa (2010, 139), who writes that "becoming plurilingual and pluricultural through life experiences and living with(in) other cultures is a complex phenomenon, which generally ends up with perceiving a change in personal identity (as a result of contacts with otherness)." At stake is the awareness that language is a cultural affair arising at the intersection of cognition and perception: "To give a clear expression of a worldview contrary to the one implied in the primary structure of a language is usually very difficult," Bohm (2002, 91) notes, "it is therefore necessary in the study of any general language form to give serious and sustained attention to its worldview, both in content and in function." Sociolinguistics speaks of *interference* in bilinguals to refer to the cognitive overlap between different linguistic systems: in polyglots, languages behave like waves that, when meeting, present features of superposition then reflected in thoughts and linguistic uses with idiosyncratic features. Most important, these thoughts and uses cannot be considered as deriving from the sums of separated linguistics systems; instead, they emerge as a completely different world-sensing.

Overall, the growing evidence of the synergies between language, culture, and mind has initiated a strand of research specifically interested in the processes of acquisition and use of a second language (cf. Jarvis and

Pavlenko 2007). This strand is certainly promising in that it provides an initial comparative avenue that somewhat puts languages (and cultures) in perspectival dialogue; however, notwithstanding the need to also conduct studies on how language is written and performed in context (beyond its normative rules), the risk is to reify a classical understanding of reality—positing dichotomies such as in/out, self/other, language/culture—where "things" have prominence over processes and where these latter are not yet considered an implicated order of emergence—as part of a whole in the making—but, at best, as a virtuous self-reinforcing circle.[9]

Recently, psychologists Liane Gabora and Eleanor Rosch and physicist Diederik Aerts (2008) developed an ecological theory of concepts based on quantum formalism. Their quantum-like hypothesis is that concepts can be considered structurally entangled with regard to both mental constructs and the given context in which these concepts are summoned. In other words, concepts are in standby, only waiting to be summoned by contexts, with language working as an actualization of thought and experience; words and sentences are, after a fashion, entangled with contextual situations in both thought and speech and eventually actions. This has deep implications for signification as a process. As Mathias Albert and Felix Bathon (2020, 440) argue, by comparing quantum and system theory as applied to the social realm, meaning becomes "fundamentally unstable as the 'actuality core' of each determination shifts again and again. . . . On the other hand, signification becomes self-referential as it provides for its own ability to be updated again." This not only calls for a discussion on meaning that foregrounds it being an ongoing process of contextual negotiation; more radically, language and culture might be said to form an entangled *meaningful* whole that cannot be studied by taking its constitutive elements apart, a conclusion that evidently requires an entirely new approach beyond mere cause–effect links. Physicist Harald Atmanspacher (2020) synthetizes this well, claiming that "in a way, the experience of meaning can thus be understood as a ('sixth') sense modality for 'perceiving' psychophysical correlations"—and this, to be clear, applies not only to any language but, more broadly, to any means of expression.

In fact, it is especially with the consolidation of the digital ecology that we can attempt what semiology has always aimed to do within and through language but could not for lack of alternatives: to reach a "degree zero" of *écriture* (Barthes 1953) and, from a hetero-*dispositif* standpoint, unveil language's sociocultural–ideological fabrication. When Roland Barthes (1972, 209) eloquently writes that "the sign is not only the object of a particular knowledge, but also the object of a vision analogous . . . to the molecular representations used by chemists," he manifests a (logocentric) position that needs unpacking. To be sure, this has already been the concern of much deconstructionist work in the 1970s and 1980s. In his *Of Grammatology*, for instance, Jacques Derrida (1976) contests the link traditionally connecting speech and writing, by reclaiming the autonomy of the latter as a form of meta-signification that relocates the object–subject dualism under the umbrella of textuality: "textuality," Gayatri Chakravorty Spivak (1976, lvii) writes in the translator's preface to his book, "is not only true of the 'object' of study but also true of the 'subject' that studies. It effaces the neat distinction between subject and object." Textuality configures, therefore, an endless sign-chain that questions the solidity of any onto-epistemological claim. Derrida's work contributes to reveal how Western thought—and its transmission—are deeply historicized phenomena, regardless of the attempt, either through speech or writing *stricto sensu*, to erase the roots of their own institutionalization. The text is but an illusory means of assertion of *praesentia*—what Derrida calls "metaphysics"—that shall be open to ongoing deconstruction. Paradoxically, considering that it is the independent nature of the fully alphabetic text that, according to Havelock, permitted the subject–object separation, writing, instead, is founded on the negation of the subject and can do even without reality. The problem with Derrida's argument is that it remains (consciously) confined within the West when, instead, such critique of presence has no intrinsic reason to be entrapped there only. As critics have noted (Meighoo 2008; Jirn 2015), logocentrism is not unique to Western civilization but to language as a means of communication. By considering, then, language as an enframing *dispositif* by default, differences in terms of textualization can be put into perspective as technocultural instantiations that demand and entice different media practices of production and fruition based on the features of each specific operating system. Being *dispositif*-dependent, every center is relative; through an ecological approach, "centrism" gets simultaneously diffused: the center also always contains its periphery.

Connections with quantum physics have also been established by semiology and philosophy as a form of legitimation of "poststructuralism's rejection of Western logos and subject/object duality, the decentering of author and authority, and the disruption of the relationship between signifier and signified" (Burwell 2018, 26). Beyond such attempts of *mise en abyme* of enunciation within and through language, today's digital ecology allows one to move out of the language ecology and look at it from the outside. In this way, the endless bootstrapping of language by deconstructionism is also bypassed to project over it an external critique, showing what a different ecology, under the spell of a different *dispositif*, can do to the language one, leading to discuss certain positions in matters of subjectivity, meaning, and experience, in terms of both their intraecological consistency (or even possible opposition between technocultural fields) and their interecological relativism (this, of course, also happens the other way around with the digital ecology explored through the lenses of language and the critique of a certain mystical understanding of objectivity and technology).

Before proceeding, it is useful to quickly sketch how the whole world has reached today's (networked) sociotechnical condition.

The Mechanization of Writing

Marshall McLuhan (1962) and Elizabeth Eisenstein (1979) both observed that the invention of the printing press (which greatly accelerated the impact of alphabetization) was at the root, in the West, of the murderous religious upheavals during and after the Renaissance; after all, as an epitomizing example, it suffices to think about the genesis of Luther's reform and its impact on Europe. The transition from the collective consciousness of the Middle Ages, predicated on the common goal of community salvation, to the new social order of public space and private minds took centuries of brutally fierce ideological and political strife.

In the protracted period of religious wars, one could see an example of the misalignment of two technocultural fields, one dominated by the oral establishment (preaching elites) and the other by the rising literate intelligentsia. Probably the institution that best represents the diametrical opposition of the old oral to the new literate subecologies was the Spanish Inquisition, inaugurated in 1478 to combat heresy. If the purpose already belies by inference the new freedom literacy was giving to individual interpretation of the scriptures, the method was even more so in that it zeroed in on the principal source of the problem, private consciousness. Indeed, the *Quaestio* required torturing the body to access the mind, that is, the privacy of consciousness. The Inquisition, proportionately to the increased readership allowed by the printing press, dismembered real living heretics in a brutal form of political control.

In a previous period of opposition between the oral establishment and the people "liberated" by literacy, Aeschylus's Prometheus, who claimed to have brought the alphabet (*syntheseis ton sumballon*, the "assembly of letters") to the humans, was punished by Zeus, who exposed him, bound to a rock, to the merciless changes of weather conditions. The myth could be interpreted as a mere symbol of power, but because it also includes the daily appearance of an eagle entrusted to devour Prometheus's liver—the source of intelligence in early Greek anatomy—that symbol also points out to a powerful "interiority of consciousness" comparable to what the Spanish *Quaestio* was up against.

The invention of the printing press is situated, in the West, to approximately 1440. This implies that in less than forty years, there had been enough "liberated" minds to require an institution to control them. By the end of the fifteenth century, the literate human was reborn, and the Spanish Church would have to resign itself to form a new generation of clerics dedicated to spiritual guidance that in France went under the name of *directeurs de conscience*.

Of course, Gutenberg was not alone in "inventing" the press. Chinese engineers had discovered and implemented wood printing long before 1440, but the invention did not prove to affect China to the same degree as it did in Western Europe. It did not occasion religious upheavals or the acceleration of scientific investigation. Insup Taylor and Martin Taylor (2014, 89) point out that for centuries, literacy remained a privilege of "a tiny upper crust of males while preventing the spread of functional literacy among the masses" (and this, as Pae [2020, 132] points out, might have been facilitated by Chinese characters not being easy to learn and master). On top of that, Pae (2020, 132) stresses that "the Confucius classics were the main subject of the institutionalized civil-service examination in ancient China," thus in a way reinforcing the replication of the status quo, rather than soliciting emancipation. In other words, not only the fabrication of logographs as an operating system of meaning but also the content they supported have hindered individuation in China.

Notwithstanding these technocultural differences and chronological discrepancies, it is undeniable that the press enabled an unprecedented circulation and cross-fertilization of knowledge. The standard test of the mechanical principle gradually superseded religious or temporal authorities as per what was to be validated. Heads were allowed to be wrong. But validity—it is worth reminding—is and has always been an effect, not an ontological principle. After all, far from depending on universal parameters, knowledge reliability and trustworthiness are determined by cultures and epochs. In his *A Social History of Truth* (1994), Steven Shapin correctly observes that each culture builds its own concepts of epistemological accuracy and trustworthiness, based on "the expectation that knowledge will be evaluated according to its appropriate place in practical, cultural and social action" (xxix). There is no in and out, before and after separating culture and knowledge when it comes to establishing—beyond a critical mass—what is to be regarded as the accepted truth.

The "Global Village"

Although the rapid acceleration of innovations in pure and applied science and technology always posed a threat to the established order, the Western cognitive norm remained relatively stable until the appearance of electricity. As we will see in the next chapter, the space-time–identity bubble of the literate cognition, separating clearly the objective and subjective "realities" of cognitive agents, has now been reversed by the advent of new technologies.

The roots of what McLuhan (1962) labeled "global village" can be found at the technohistorical intersection of the explosion of (electric) mass media and the global order following World War II, which rested on two opposing blocks—US and USSR—and on economical–political agreements—Bretton Woods—as well as on the establishment of the United Nations and the Universal Declaration of Human Rights.

In the years following World War II, the West reached its peak of industrialization, a process that went hand in hand with mass laborers' movements and parties. Concerning media, the two decades after the end of the war represented the golden epoch of TV, which rapidly became the dominant medium in citizens' information diet, overcoming the press (which still relied on rigid publishing processes and schedules) and the radio (which could not compete with the visual imaginary built by TV).

In the late 1960s, an economical shift began to impact heavy factories in favor, especially in the West, of tertiary services. The media ecology was not excluded from such changes, witnessing the "liberation" of radio stations, commercial TV channels, videos on demand, and portable music devices. Socially speaking, the boom of consumer society enlarged middle classes as a consequence of economic prosperity, but these also tended to get increasingly diversified, according to the redefinition of transectoral needs and civil–political sensibilities. The (fabricated) universalism that permeated the early aftermath of World War II got rearticulated into particularities whereby the normative, hegemonic conception of political space was reappropriated by either indigenous stances (e.g., postcolonial movements) or individual agencies (e.g., feminism). Or both.

Politically speaking, it was Margaret Thatcher's idea that "there's no such thing as society . . . and people must look after themselves first"—mirrored by her counterpart on the other side of the Atlantic, Ronald Reagan, who claimed that "government is not the solution to our problems, government is the problem"—to mark a milestone in the gradual decomposition of society-as-collective (this also applies, in part, to Deng's Xiaoping reforms in the 1980s, which broke with Maoism, ushering China into a new era). Such a process can also be detected in the media ecology, with increasing market deregulation of broadcasting spaces and the selling of public assets to private corporations. It is not by chance that between the introduction of the personal computer (PC) (1984) and the advent of the Web (1991), the major sociopolitical turning point of this epoch occurred: the fall of the Berlin Wall (1989). The fall might be said to have been "pushed" by the digital–electronic revolution that PCs and telecommunications preconized by wiring the world on a global scale, alongside the full deregulation of the media ecology put forth in the mid-1990s, first in the US and later in the rest of the world.

The Web

If the origins of the internet—literally the "network of networks" that supplies the infrastructure for the Web—are to be traced in the mid-1960s early 1970s, as the development of military technologies for encoded communications during the Cold War, the Web was invented and developed by the team led by physicist Tim Berners-Lee at CERN. Still in its experimental form, the Web was officially launched in June 1991: Tim Berners-Lee and his team developed the first software that allowed two computers to communicate and, right after, the hypertext markup language (HTML) and the Web browser that allowed to solve the problem of "translating" and distributing text documents across computer systems. It was then in 1993, after two years of testing and optimization, that the Web opened to the general public.

The ecology of text before the appearance of the Web consisted of recorded or printed textual objects distributed in space, requiring classification, storage, and recycling. With electricity and digitalization, text mutated and became hypertext, further leading to hypermedia and virtualization. Hypertext is ubiquitous and accessible instantly on demand. It has modest storage needs (by comparison with text), but it requires complex access routines. As the main support of language shifts from the printed to the electronic word, text is redistributed and paradoxically resensorialized in multimedia.

In the early 1990s, envisioning the huge impact that the Web was going to have on society and people's psychological understanding of self and collectivity, de Kerckhove (1997a, 189) noted, "Not too long from now, our sense of belonging to a common ground will come principally not from TV nor from physical space, but from our connection to the Internet or whatever the Internet presages." The most interesting argument is that media such as cinema, TV, and PCs are not only technological apparatuses but also *psychotechnologies*, that is, artifacts that affect humans' psycho-cognitive faculties. While cinema and TV are able to foster a sense of collectivity (these are de facto one-to-many mass media, whose fruition is framed within precise locations and times), the increasing portability of PCs and digital devices has dispersed such collectivity, gradually metamorphosing it into an independently connective system. Questioning this paradigmatic shift, in terms of the psycho-communicative empowerment *and* disempowerment that the PC and especially the Web bring to the fore, de Kerckhove (1997a) asks, "It is not quite so clear what we are going to do with all this communication power. How do you change the habit of relying on the automobile for power, action and prestige to the adoption of 'telepresence' as a way of being?" Now we clearly see the answer to this question. Of course, it is not only a matter of readapting literacies to the digital age—studies on that have been conspicuous (cf. Hayles 2012; for an overview, Delgado et al. 2018)—rather, at stake is a matter of attitudes, orientations, and, most important, the role of language as a meaning-making *dispositif* in the digital age.

This is especially the case today that generative AI can produce language with minimal input from humans. With a few words as prompts, digital wizardry can create poetry "in the manner of . . . ," redact scientific look-alike documents, program functioning software, make music or pictures, and many more benefits—and perils. Mechanizing language amounts to taking control of it and, with this control, managing humans as Stanley Kubrick somehow predicted with HAL, the operating system of *2001: Space Odyssey*. What so many people fear today is that AI will take over the quasi-algorithmic power of language colonizing humans to serve automated interests that have little or nothing to do with human aims. Notwithstanding the alleged "goodness" of human aims, we do not share this fear, having observed and studied several instances of technologically driven social change and having concluded that, after the predictable strife and disruption each transition entails, human societies have eventually recovered an equilibrium. The point is under which *translating conditions*. What is at work today under our very eyes is the rapid hybridization of one set of operating systems, those of language, with another, that of AI algorithms. And that will be the topic of the next chapter. But the digital ecology currently on its way is only a prelude to yet another disruption, that by the quantum ecology.

2 Epistemology: Language and Data Superimposed

> Today's digital age is beset by an information crisis that revolves around the five giant evils of confusion, cynicism, fragmentation, irresponsibility, and apathy.
> —LSE Truth, Trust & Technology Commission (2019)

One of the most appalling aspects of the pandemic that struck the world in 2020 was the states' inability—or even reluctance—to coordinate a global response by acting collectively. Especially during the early stages of the outbreak, governments entered bilateral accusations, engaged in reciprocal blaming, and, after being pulled inertially into the crisis rather than proactively tackling it, undertook independent courses of action, often with limited vision and scope. It was only on March 11, 2020, almost two and a half months after the presumed outbreak of the Covid-19 epidemic in China, that the World Health Organization's (WHO's) general director Adhanom Ghebreyesus declared the coronavirus a pandemic. On that occasion, Ghebreyesus also denounced the "alarming levels of inaction" of various countries around the world, unable or unwilling to follow WHO's guidelines. And yet, during the peak of the pandemic, the WHO itself was subjected to harsh criticism by some of its member states, including the US, in the figure of then president Donald Trump, who overtly accused the WHO to have underestimated the pandemic and be China-centric. The WHO, differently from other bodies of the United Nations, does not have the power to promulgate binding resolutions for its member states and, as a consequence of the 2008 economic crisis, has been gradually disempowered and delegitimized, especially at economic and political levels. As reported in *The Guardian*,[1] *The Lancet*'s editor Richard Horton puts this down very clearly: "WHO has been drained of power and resources, its coordinating authority and capacity are weak. Its ability to direct an international response to a life-threatening epidemic is non-existent." Beyond the mistakes the WHO can be held accountable for concerning the Covid-19 pandemic, its lack of authority and leadership, alongside its increasing politicization, are all evident signs of the emerging multipolar world in which the compass no longer points to global cooperation but bilateral agreements: "All the previous rules about global norms, public health and understanding of what's expected in terms of an outbreak have crumbled,"[2] declared Lawrence Gostin, director of the O'Neill Institute for National and Global Health Law. In other words, the pandemic has painfully unveiled the shortcomings of globalization as an exclusively economic–commercial process, rather than a truly internationalist cooperation. The WHO crisis is just another symptom of the resurgence of nation-based politics under different guises. The global paradigm that emerged out of World War II shows internal decohesion, and the causes of that are, all at once, economic, social, and technocultural. What has made this condition unique and worrisome is that such decohesion has happened in a void of political leadership, with the US withdrawing from this role and China still refraining from truly opening itself toward the world, beyond commercial and economic advantages. And yet, while nations continue (or return) to be the main frames through which to define political and economic agendas, today's challenges are worldwide and risk to remain systematically unaddressed. Breakdown occurs at all levels, micro (individuals and local communities), meso (states), and macro (continental and international relations). With climate change hanging over humanity's destiny, this situation signals the preconditions for a potential annihilation, or at least deep rethinking, of human civilization. At the conference "Art Meets Science and Spirituality in a Changing Economy: From Fragmentation to Wholeness," held in Amsterdam in 1990 and hosting, among others, scientists, spiritual leaders, and economists, physicist Bohm wisely said, "The greater power we have, the more coherence we need."[3] Today, despite the huge power coming from science, big data, and the digital transformation, the global society has been shaken in its foundations.

Digital Ecology in the Making

As seen in chapter 1, the language ecology that led ultimately to the emergence of the global village—however differently articulated in various technocultural fields worldwide—has been gradually undermined by the acceleration that the internet and the Web, as the networked technologies *par excellence*, impress upon the whole world. Terrorism and global epidemics, even more than the specter of nuclear war, establish a running and pervasive background of enduring tensions, which show the inadequacy of an ecology torn apart by short-term and short-sighted interests and behaviors, as well as opacity and contestation among political actors and institutions. But these are just symptoms. At the heart of the tensions is the increasing misalignment between the language and the digital ecologies: frictions between their two incommensurable *dispositifs* generate—as we will discuss—a deep *epistemological crisis*, profoundly disrupting how people (try to) make sense of reality. It is not that algorithms—the operating systems of the digital ecology—contradict or oppose traditional literacies, but they contest language meaning-making logic, creating new performative configurations that elude the possibility for ongoing sharedness, leading ultimately to social decohesion.

Algorithms do so by bringing forth a redefinition of the three coordinates of space, time, and identity, as also examined by Michalis Vafopoulos (2012), who speaks of "web being" to refer to all and any "data object," thus allowing to distinguish the experience of space and time from the technology's perspective as opposed to that of the user. According to Vafopoulos, the characteristics of data objects are digital, virtual, aspatial, discrete, indivisible, recombinant, nonrival, and infinitely expansible. Let's dig into the matter.

Space

The digital spatial dimension is that in which and through which data-driven technologies act. In itself, such a space already constitutes a form of deconstruction, or at least dematerialization, of physical space. With the emergence of metaverse technologies, the space of digital technologies can be dubbed a virtual doubling of physical space. And yet, it is a doubling that is enmeshed with physicality, which gets, indeed, increasingly virtualized. It is then a doubling that generates a hybrid, a techno-psychosocial ensemble. Thus, since the invention of the internet, humans consciously occupy three distinct spaces. When connected to the internet, instead of occupying only one's own physical and mental spaces, one also inhabits cyberspace. These three distinct spatial dimensions are occupied centrally: the physical, material environment (including the occupation of one's body); the mental, cognitive, privileged milieu of imagination and thoughts; and now the total network of digital space.

People enjoy—and take for granted—an enormous increase of personal powers. As they sit in front of a computer, users occupy a huge environment of information-processing and decision-making just as they do mentally and physically, but in vastly expanded ways and with far-reaching operative tools. This new digital realm finds a place between the physical and the mental worlds: indeed, it is partly physical, by occupying an intermediate virtual realm, and partly mental because it entertains and shares an interactive cognitive relationship with the user's mind. Vafopoulos (2012, 412) calls this intermediate condition "aspatial": digital beings are aspatial "in the sense that they are everywhere and nowhere at once." Entering and negotiating digital space requires different kinds of interfaces, besides body and mind. Although the digital realm is outside bodies and minds, it is increasingly synchronized and tuned to tech devices to such an extent as to not distinguish neatly between the three spaces. Also recognizing the spatial novelty of the internet, Salvatore Iaconesi and Oriana Persico (2015, 140) have articulated this new spatial experience in a focused observation:

> We find ourselves in a digital Third Space, more inclusive, in which information is not only attached to places, spaces, bodies and objects, but constantly recombines, remixes, recontextualizes, creating constantly new geographies which are emotional, linguistic, semantic, relational, or relative to the many patterns that non-human algorithms can glimpse in the way layers emerge from data, information and knowledge, correlating different spaces, times and human networks.

As the two authors note, the interfacing of the three interpenetrating spaces has given rise to a reconfiguration of the techno-psychosocial landscape, not only as a recoupling of factors but as a new emerging dimension. This is the cyberculture, which is the result of the multiplication of mass self-communication (Castells 2007) by instant speed. Cyberculture implies "seeing through": seeing through matter, space, and time with information-retrieval techniques. Whenever a technology gives people mental and physical access to somewhere on the Earth or deep in space, beyond any previous limit, their minds follow. Hence, psychology must evolve with that technology. People take for granted that GPS tracks and guides their itineraries, but whether they travel on business or for pleasure, they are physically, mentally, and virtually expanded in the global sphere. When people think globally and send or receive messages on their mobiles, they contain the Earth in their minds and in their networks. The information applied to this inner structure is part of a global thought and a global activity. As a form of expansion of the mind and frame of reference, networked globalization is, then, one of the psychological conditions of the digital transformation. The convergence that is produced, indeed, is not only a convergence of contents and media (Jenkins and Deuze 2008) but an epistemological convergence, a convergence of formats.

The digital transformation, if anything, transforms the experience of space (and time), returning them to the complex unity that general relativity and nonalphabetic cultures tend to associate with them. Vafopoulos (2012, 420) specifies that "the Web 'curves' physical time and space by adding flexibility, universality, more available options, and sources of risk." Now, however, the multiplication of mass and acceleration of speed has pushed global connectivity to extremes that provoke a kind of reversal, tearing gaps in the cybercultural tissue. While functioning as a social limbic system carrying emotions across geographical and cultural frontiers, the internet has also instilled a surreptitious rationalization and individuation of /reality/. The incipient digital transformation crept in with promises of "disintermediation," that is, freeing consumers and citizens from the constraints of established business and administration practices and dicta. What even the occasional naysayer had not predicted was that the digital transformation has introduced yet another level of technologically based abstraction. The effect is to further isolate individuals from their peers in a comparable but more efficient way than the technologization of decision-making by the bureaucratic state has done (Bauman 1989). This is the origin, today, of much hateful debates online (and increasingly extreme behaviors).

Hence, digital experience constantly redefines communities at a virtual *and* physical level, molding and remolding a sense of belonging that is sometimes local, loosely international, mostly transliminal. No longer just a matter of merely participating in cyberculture, it is cyberculture itself, as a node culture, that defines and positions its actors: as Ulises Mejias (2013, 31) notes, "What we are seeing is not only the pervasive application of the network as a model or template for organizing society but also the emergence of the network as an episteme, a system for organizing knowledge about the world."

Time

The digital transformation has introduced an experience of time as simultaneous and instantaneous, an effect that tends to put narratives in question. As François Lyotard (1984, xxiv) perceptively put it, postmodernity is characterized by "incredulity towards metanarratives." Long, historical, almost teleological narratives have shattered and been somewhat discredited. This is an effect of speed. McLuhan (1989, 34) observed that

> electric speed tends to abolish time and space in human awareness. There is no delay in the effect of one event upon the other. The electric extension of the nervous system creates the unified field of organically interrelated structures that we call the present age of information.

Now, such extension is being reconfigured according to fluxes and fields that are as fast as electricity but also increasingly dematerialized: it is a "timeless time" (Castells 1998) in the making. This timeless time is occasionally understood as "real time." The experience of real time contributes to a global opening and consequential imploding bubble. The narrowing gap between action and reaction creates a kind of continuity between planning and executing in real time, which makes the timeliness of action necessary, but also its performative decay very rapid. Real time is a meaningless time (much different from *present* time); it is a flat time in which there is no opportunity to reflect in the mirror of one's conscience. In fact, the only reflection is that which comes from the screen, a kind of fairytale mirror that reinforces the fiction the viewer is enslaved by rather than a tool for (self-)understanding. As Berardi (2015, 13) notes, with/through digital technologies, people's psycho-physiological responses have reached a point of "exhaustion."

Unless structures in place can bend or resist, the consequence of acceleration will disintegrate or mutate them. The acceleration of water molecules by heat turns the original substance into vapor. Likewise, in no time, acceleration will push to the brink established industries and institutions. This is the first effect of what people call disruption. Sudden technological and social acceleration without preparation can indeed bring giant corporations to their knees. And it can happen in a matter of days, reconfiguring the technocultural landscape of which these actors are one among many gears. Just think about how ChatGPT has forced Google to rethink its long-lasting hegemony and business model. By getting accustomed to it, most people accept speed and constant acceleration as the new normal. Suffice to think of the hardly surprising difficulty, if not impossibility, to slow down or reconvert socioeconomic paradigms in times of crisis. Computers have increasingly accelerated psychological responses to the point of breakdown, much faster than planes, trains, and automobiles ever did. And they did so by removing the need to change places.

Situated in a *u-topic* ("nonplace") dimension at the center of things, people do not need to move anymore, if not for leisure and distraction, or for the sake of covering distances. Smart-working during the pandemic is proof of that. This means that instant access to information makes people not only *prosumers* but also *prosumed*: their actions are marked by their personal character, and in turn they become the perfect target of immediate and constant info-bombing. Computers and data mining have synchronized the global network as waveforms of electrical currents in electromagnetic fields. And they did so by virtualizing space and redesigning the physical nature of information. While a one-way, frontal relationship with the TV screen ushered in mass culture, computers and smartphones, introducing two-way interactive modalities, have added speed. TV created the notion of the "mass subject," while computers brought in the "speed culture": the "speed subject" of computers, then, is everywhere at the center of things. The computer is not a mass medium but a personal one for global instant messaging.

Every time the emphasis on a given medium changes, the whole culture shifts. Computers induce an adrenaline "doped" biopsychological state of alteration (just think how difficult it is to fall asleep right after turning off your laptop at night), because computers, differently from TV, which is an absorbing medium (like cinema), respond back: they demand cooperation even when users adopt a passive stance toward them. It is no longer (only) a media experience; it becomes media devouring, and yet—to be sure—it is not the user who devours content; it is the medium that devours the user. In a recent article, Martin Korte (2022) has reviewed neuroscience research on the effects of digital technologies' use on the brain, as well as on cognition and behavior, and he has found that, although more experiments beyond self-reporting are needed, research does highlight tremendous impacts of these technologies on, among others, language development and the recognition of emotions.

The deep involvement required by digital watching and the fact that most responses are (almost) involuntary bear witness to the changing power relationship between consumer and producer. When people read, they scan the text; they are in control. But when they watch screens, they are being "read." When scanning meets glancing and makes eye contact between man and machine, the machine's glance is the more powerful. And yet, while with TVs people's experience is still framed within a domesticated environment, nowadays portable digital devices have extended into public space, also exteriorizing such power asymmetry and involuntary responses. Suffices to think of those people who, when facing situations of imminent danger, decide to take videos of it, thus adopting an unreflexive response, instead of intervening or running away. The spell of technology has come to subvert even basic surviving principles. Screens crave for humans. As viewed from the technology's perspective, the basic unit of time is attention. This commonsense but not obvious observation brings back Henri Bergson's (1965) insights about duration, with the key difference that now such duration is no longer private or qualitative but externalized, that is, discoverable, observable, traceable, and processable (Vafopoulos 2012, 408).

Identity

The present era suffers from a global crisis of identity. No matter what culture or creed, every human today is either directly or indirectly affected by the threats the digital transformation is posing to individual and social forms of identity. There are several reasons why people lose their private and collective identity, even as they regroup into unevenly connected communities of interest to find one.

First of all, the real object of digitalization is to extend to the electronic environment the kind of control and monitoring relationships people experience between them and within themselves. Digital technologies effect a kind of increasingly embodied social mediation in a single continuous extension and externalization of individual powers of imagination, concentration, decision, and action. This phenomenon, following and crowning a durable French intellectual current beginning with Leroy-Gourhan and continued by Derrida and Foucault, Stiegler (2015) calls *hypomnesis*, that is, memory, and by extension cognition, occurring outside the mind, as opposed to Plato's *anamnesis* that locates them within.

These technologies function largely like a second mind, one soon to be endowed with more autonomy than users might care for. Thinking emigrates online in AI software and supports the expectation of the "augmented mind." Indeed, the very concept of augmented mind indicates that instead of merely occupying a neutral physical space that communications "traverse," people occupy a cognitive space that contains them and which communications bind and support.

The stability that literacy provided by favoring personal cultivation and expression on the public stage has been shattered and polarized, atomized even, to the point where people no longer trust their own assumptions. Digital technologies have emptied the self from within, externalized its faculties, eroded judgment, spoiled meaningful sharing (another word co-opted by the digital ecology), and eventually fueled (self-)distrust.

More forcefully than his predecessors, Stiegler repurposes in political terms Plato's famous *Phaedrus* argument about the loss of internal memory to writing. Years before the onset of generative AI and large language modeling (LLM), Stiegler (2016) predicted the radical takeover of language itself by digital technology:

> Digital technology is a form of writing, a writing that is produced at the speed of light, through machines to which we have delegated the process of reading and writing, organized and controlled by a planetary industrial sector established by global companies that have been in existence only a very short time. Digital reading and writing constitute the new milieu of knowledge, in fields as diverse as astrophysics, nanophysics, biology, geography, history, mathematics, linguistics, even sports science. Therefore everything, absolutely everything, is in the process of becoming digital. We are witnessing a total mutation of knowledge, which affects at the same time embodied knowledge and life wisdom. Daily life is what is first upended, in all its dimensions.

While everything that constitutes personal identity—memory, intelligence, imagination, judgment, and privacy—emigrates outside to databases, users' minds are filled by contents that are pushed into them by algorithmic—and occasionally caricatural—representations of what people *might* need or desire. Such representations are not really the subjects' identity, but they serve as a format that people inform (and by which they are informed) at the cost of not really developing discernment, self-reflexivity, or who they could become. A recent dramatic demonstration of this was given by the documentary *The Social Dilemma* (2020), which demonstrates the way school-age boys and girls are visibly and effectively manipulated for commercial gain.

This is the "digital unconscious," that is, everything that is digitally known about you that you do not know. Like Carl Jung's notion of the collective unconscious, it is founded on ambient but latent information.

But, and that is key, the digital unconscious is—at least in theory—a private one. The collective unconscious is supported by stories people tell, retold in documents and present in vestiges that distribute more or less evenly the collective memory of myths and archetypes that occasionally emerge in consciousness. The digital unconscious is present in a growing quantity of databases and guided interactions that are managed by ever more intelligent software to address individuals, not only collectives. The digital unconscious differs from the collective one by the immanence, permanence, accessibility, and potentially instant and global diffusion of data, instantly available for collecting and reconfiguring to emerge at a conscious level in real time. The digital unconscious is borne as much by the personal data and network activities that people are more or less aware of, as by automated referencing systems that can offer tailored data about professional activities, travels, banking, and medical and other records, including all the social traceability. This has ushered humanity—willingly or not— into an era of transparency. Augmented, instantaneous, and omnipresent, the digital unconscious introduces pervasive transparency, both personal and collective. But transparency also requires equality, a new symmetry. If one can be so deeply, immediately, without protest verified as an object of observation, absolutely the same transparency must be demanded from the system. "The thing to fear," Joshua Meyrowitz (1999, 111) said, "is not the loss of privacy per se, but the *nonreciprocal* loss of privacy." Hence, the era of transparency is also the era of *suspicion* (to borrow the enigmatic term invented by French author Nathalie Sarraute), as the concretizing of whistleblowing practices attests.

As a consequence, an atmosphere of latent anxiety emerges. There is no defense against the growing *suspicion* that, yes, bad things could happen here and now, to anyone, under one's nose. From this precondition (re)emerges an ideological vision that privileges a reactionary conception of the individual and society. And yet, the masses that characterized the populisms of the beginning of the twentieth century have in the meantime washed down. Today, what remains is a populist feeling, one that is disenfranchised from the people as a collective and connotes itself for advancing a precarious "glocal" (tech-mediated) identity within the increasingly uncertain context of globalization. Slogans such as "America First" and "Out of the EU" are symptomatic of that, as well as individual attitudes labeled as NIMBY ("Not In My Back Yard"). No surprise, then, that social media are by

now polarized spaces, lacking and preventing any form of collective aggregation. One could also suggest, in biological terms, that it is the genotype
of the algorithm (which is binary and made of bits) that has come to affect
the phenotype of human thought (which is syncretic and made of atoms).
Binary formatting is indeed the precondition of the digital ecology to create
a tabula rasa of the "sensible," which becomes, in this way, ripe for computation at the cost of its reduction in richness, diversity, and complexity. As
an example of the impact of the digital on the language ecology, it suffices
to refer to cognitive research (cf. Pashler 1994), which has robustly shown
that whenever linguistic activity is interfered with numerical activity (e.g.,
naming vs. counting), people's reaction time becomes much slower because
the two tasks attach to different cognitive mappings. In these cases, people
literally become "dumb."

An effect that is complementary to transparency is that especially online,
one can select to be anonymous or pseudonymous, if not always to the
tracking capacity of digital activity, at least to the users engaged with, for
whatever purpose. This, along with the gradual softening of judgment,
allows anyone to eschew responsibility toward anyone. In physical space,
one may be held to community, family, or oneself but never to the same
degree in virtual space. Online, one has no shame or guilt feelings. As Ivo
Quartiroli (2011, 180) puts it, "In the post-printing press culture, the ego
personality has become fluid, but only apparently more open. Exposed to
greater volumes of information that all compete for our attention, we are
split by multiple inputs without a director to integrate the personality at
the center of consciousness." Such identity dispersion is not surprisingly
disorienting and poses serious threats to collective responsibility. Through
the internet, what has emerged, instead of auspicated harmony and democ-
ratization, is epistemological divergence.

The Renaissance witnessed a brutal redefinition of what was meant to
be human during the transition from a predominantly oral and communal
religious authority to an individualistic social and political order. While
transiting from shame to guilt, the object of personal responsibility in Western societies (not so much in Asian societies) shifted from the other to the
self. Today, as people are ever more exposed to continuous monitoring by
automated electronic systems, responsibility shifts away from the self to the
almost self-organizing whole social order. This shift attunes more to Asian
societies: it is indeed quite possible that the sense of honor sustained by

the fear of "losing face" will return as foundational, if only to protect one's "reputation capital." A new feeling, neither guilt, nor shame, just a mild but persistent confrontational attitude, has emerged. More people now know that they are being tracked for everything (to which they sometimes naively reply "to have nothing to hide"), but they have not yet understood that the invasion goes far beyond exploiting personal data entered in Meta or disclosed by X; this invasion literally wires to subliminal, survival behaviors. As McLuhan wryly observed during one of his classes, "The more they know about you, the less you exist."

To be sure, the main problem is not solely related to the ethical use of technology, although this is certainly worrisome. The fundamental problem is that, as usual, every major technological shift changes the (psycho)-social ethics, as well. Whenever the medium that carries communication changes, it introduces changes in social and personal behaviors. Through electric technologies, the ethics of the individual person becomes that of the *enforced* social person, or better of the mask, as the fabrication of the social subject. Electricity connects and opens all barriers but also isolates and uproots individuals, whereas literacy has created divisions and categories but also supplied the ground for commonality.

The Fabrication of Meaning (and Meaningfulness)

One way to understand and assess today's epistemological crisis[4] is to expand the understanding of what an algorithm is. An algorithm can be understood as a set of instructions to effect a command or exercise a control between communicating entities. The algorithmic function per se is not exclusive to digital systems. For Vico (1744), commands and modalities of social control began with gesture, then proceeded to sounds, became codified as language, and were finally written down. All these modalities require the intervention—and the negotiation—of meaning. A cry of pain is an instance of communication that needs interpretation from the listener, animal or human, where it stimulates at least a reflex of mere recognition. So, one could consider language as a loose algorithmic system, in that it processes human activity in a complex and reliable fashion; it functions as an interoperable system for making sense (input) and acting (output) upon the world.[5]

Eventually, it is when commands get translated into digits that they cease to require/embed meaning. The disappearance of meaning as an indispensable intermediary for critical operation, decision, and execution of the command chain erodes intersubjectivity and brings an entirely new form of intentionality. This underlying transformation of humans' customary modes of expressing intention, still mostly hidden to awareness, is a major feature of the epistemological crisis humanity is undergoing right now.

The Crisis of Meaning

This profound crisis of meaning is reflected, among other things, in the chaos of the "infodemic,"[6] as well as in the challenges to science and in the negation of objectivity, largely effects of the digital transformation in progress.

Statements online can be transmitted to thousands of people on the basis of a vague intuition or a tenuous relationship between the signifier, that is, the statement, and its social context. The epistemological crisis begins when too many people no longer distinguish the difference between objectivity and subjectivity or give priority to subjectivity without any other form of proof. It goes without saying that individual judgment suffers from this disconnection from the "reality" of statements: politicians know this way too well and capitalize on that. Governance by tweets has enjoyed unlikely success, regardless of its inaccuracy, and so does the flood of fake news for fun or profit. The latter has become an industry with production, delivery centers, and specifically hired people to do the job. Getting the truth from the internet requires will and skill. What is said in fake news may not be supported by facts, evidence, event, or even common sense, and yet, the speed of the transmission and the lack of contexts to duly question and debate what is presented bypass or hijack any form of certification, so that shock or assent and consequent emotional responses become the default mode for coping with information. This fluidity and evanescence of factuality make any claim worth considering as long as it hits the right buttons of anger, indignation, fear, or derision.

The crisis of meaning is now explosive, not only because people take their desires for given but also because at the highest level of human affairs, algorithms are replacing classic rationality in making decisions. It is not only that the meaningful context disappears from the mind-set of less educated populations, but meaning itself is sidestepped by algorithmic logic. The epistemological crisis is contemporary—and perhaps complementary—to that of meaning in global culture. There is a dangerous indifference—probably more in Western than in Asian cultures—to the obvious, to rationality, to objectivity, and even to the simple need for common references in the world. And since the crisis of meaning already finds citizens caught in a negative socioeconomic spiral, all this leads them to resign themselves to trust digital technologies more than their own judgment. It is hence in the algorithm that the principle of objectivity, without which a valid decision cannot be made, ends up taking refuge. Judgment thus passes definitively from the head to the machine.

The Loss of Referent (and the Public Sphere)

The referent is the third side of the Saussurean triangle of the sign in figure 2.1.

Many interpretations of Ferdinand de Saussure's seminal study ([1916] 1959) of meaning are available. Figure 2.1 has the merit of starting at the left with a road sign and not with Saussure's academic term for symbol, the *signifier*, often confusing for the uninitiated. Anybody, in any culture or language, can recognize a road sign, the meaning of which jumps to mind spontaneously. The purpose of the "No Parking" sign is to make you think on the spot that you cannot park there, and that is a mental interpretation,

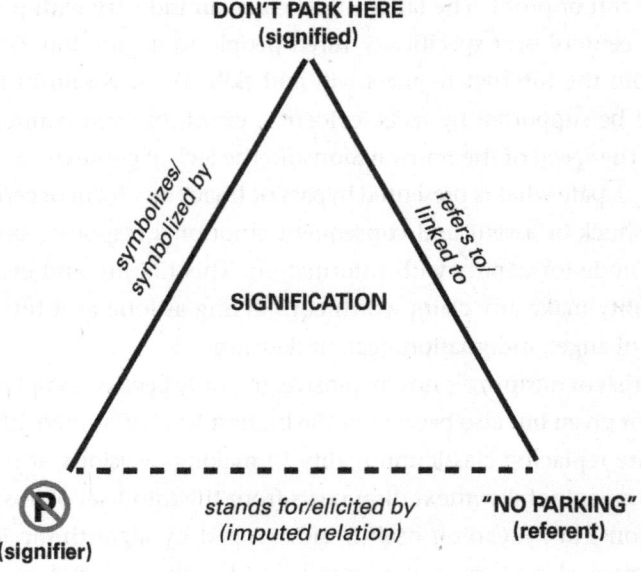

Figure 2.1
An example of the triangle of signification.

otherwise called the *signified*. The job of the material sign is to generate that mental response. So far, so good. And yet, why is there also the need for a *referent*? Because the recognition of what the sign means is not enough to complete the cognitive operation of signification. In other words, you also need to know that "No Parking" is not some arbitrary fiat but an existing rule in the world—a normative prescription that is part of the traffic laws, which is constituted as an agreed-upon reality independently from that particular moment of recognition. In fact, the line between the symbol and the referent is dotted, suggesting that it is a connection that is not given but needs to be repeatedly confirmed in practice (it is the referent as a *tertium datur* between object and sign that second-order logic fails to recognize). Note also that such connection is considered true, implying that the purpose and very presupposition of that relationship is to codify for the collective—that is, in a shared way—the veracity of the meaning of the whole sign. The genius of Saussure was precisely to add the requirement to verify the validity of the symbol–reference relationship, at least in one's own mind and memory, if not going to the source (not really necessary in the case of a real road sign because its presence in front of your eyes is its own referent, but much more relevant in an infodemic context).

If Saussure's triad is intended to describe how language relates to meaning, it works well for alphabetically written language because words (except for a few imperatives and declarations such as "yes" and "no") are not and cannot be at the same time the meaning and the thing they say (another thing is the performativity of language, i.e., the effects it produces). This, however, is not the case for the image (and this is also why the loss of the referent within logographic systems, where the relation between meaning and shape of the characters is not arbitrary, is different). In the image, the referent and the signifier are one and the same. The image first means itself, not something else, and only upon reflection can its meaning be expanded to interpretation. This is not some sort of naive ontology; it is consolidated pragmatics. Of course, there are categories of images, such as direction panels, that behave like words and point to referents. However, most images are polysemantic, behaving as direct evidence of what they represent. Thus, the condition of sense-making that images propose is that the referent can be ignored or absent.

In this respect, the problem is epistemological, by which we mean that it addresses the basis of how people make sense of the context they find themselves in. The loss of the referent spells the end of separating "the knower from the known" as per Havelock's foundational insight about an important effect of the alphabet. Objective and subjective come together, eliminating the separation. It seems as if humanity was returning to *doxa*, the Greek word for opinion, where only what you believe counts as real. Fake news is, in fact, the by-product of a dimension that can no longer be analyzed objectively (there is no longer a shared referent, however contested it may be) and a phenomenon that can only be discussed in terms of probability (both fake news per se and their content).

To be sure, the referent is not something given once and for all; it is always a sociocultural construct. And yet, the referent is a communally negotiated idea, one able to catalyze and stabilize agreement among people over time. Now, this is less and less the case. While people create and share consensus on the referent, algorithms skip such a consensual process; when they perform actions such as automating driving or legal decisions such as what is practiced in "legal tech," they enforce solutions in such a way that meaning and meaningfulness are not involved or needed—it is simply a hetero-directed enforcement. AI carries this process even further by configuring "simulated meaning" (such as deep fakes or generative AI answers to prompts) that are charged with meaning but only to the receiver from the outside, not for the sender, from within. Digital highways are merging into a single cognitive environment where the world outside is neither fixed nor real in the conventional sense but behaves variably, both synchronically and diachronically. This is a condition of enduring overlapping states, of blurring lines and crossing boundaries. A few examples concerning the fate of the referent in the digital realm:

- *Absence of referent*: Prior to a search or a question, there is nothing "out there on the Net," that is, nothing related to anything in particular in the umpteen databases storing data for further use, at least from the point of view of the user (what is not generally known is the extent algorithms behind a search engine such as Google are already able to predict what users are about to search). Hence, until the question has been asked, the data are not yet correlated and yield nothing. True, the material related to the answer might preexist in virtual space, but before it has been solicited by the question and cross-referenced, it is simply not there. The space of the Web is somewhat metaphysical, or at least paraphysical. What algorithms do is to impose certain ranking criteria on the given response, thus creating a convergence toward a ur-referent. Furthermore, being motivated by need or desire, more than by a constraining context of referents, big data questioning can produce entirely new contents, just as the mind can produce fiction, an aspect exacerbated by generative AI such as ChatGPT and its growing number of variations.

- *Extracted referent*: This is the case when a search engine powered by AI studies a large body of synchronic or diachronic data—say, a substantial corpus of historical documents—and can dig out in the analysis trends, word frequencies, and recurrent coincidences, as well as reveal unsuspected linkages or patterns about the appearance and the development of concepts or innovations. There appears in such cases referents that could be said to preexist the search but that no one had thought about before. There is no direct one-to-one referent but a derived (inferred) referent as statistical analytics.

- *Excess of referents*: People occupy digital space, mostly unwittingly, by numerous profiles that correspond to different configurations of their various data locators. The single referent is "my-self," of course, but the data collected are assembled for a multitude of referents that are tailored to suit the needs of whatever enterprise, administration, or government, most of which usually have access to more information about them than users have themselves.

- *Referent as a leader*: In this case, the referent is only approximate and depends on a statistical guestimate, for example, presenting a keyword-based map of opinions expressed by the tweets of a given community, say, in the course of sentiment analysis. Under the general name "emotional traffic," it is obvious that such estimates are approximate at best and irrelevant for a serious study of population moods, but they are started with a credible leader in keyword probing.

- *Substituted referent*: This is perhaps the most pernicious of all fakes and hence has deserved the definition of "deep" fake. It uses photography (for its power to generate the illusion of reality), morphing, and computer-assisted seamless suturing to place—say, one person's head, or adding another person's voice and words—to a different person's body. Examples abound, ranging from silly but entertaining pranks all the way to gravely threatening fabrications putting people's reputation and credibility at risk by picturing them in compromised situations.

- *Simulated referent*: Emerging last but not least is the text, image, video, or computer program that is summoned by prompts in LLM and image, sound, and software creating systems. Such productions are somewhat "fake," in the sense of fabricated, but they are also original, getting close to what might be called "machine thinking," even though, of course, it is not—yet. True, biological and machine content production shares the features of, among others, spontaneity, instantaneity, and meaning-making, but the referent is not fact, but similitude, also based on shared knowledge, and all the more deceiving for that matter and, for that reason, already much feared by meaning gatekeeping institutions. (Note also that LLMs can speak of themselves, somewhat autopoietically, but can they also break their own rules? We might not even get to know the answer to this question or it might become irrelevant, but the *effects* LLMs have on cognition are worth exploring.)

*   *   *

With the disappearance of the referent, weakened and eroded is also the public sphere, that ideal dimension in which people can converge in order to dialectically debate on, and eventually dis/agree upon, any given matter of contention. The disappearance of the public sphere is, in fact, another perspective from which to look at the transition from the collective to the connective paradigm, caused by the networking of society. If the mechanization of writing constituted the premise for the coming to being of a public sphere—indeed, Jurgen Habermas (1989, 30) specifies that, in the beginning, such a sphere was strictly connected to "the world of letters"—the networked society has disrupted it. Major communication theories still assume, somewhat uncritically, that both traditional and new media are able to give shape and inform the public sphere. However, such an assumption largely overlooks the extent to which the public sphere has been torn apart by the nodal polarization promoted by/through the network.

The disappearance of the public sphere has causes and effects of an epistemological nature. Through the network, many users have reached a condition of mere assimilation of information (rather than acquisition), a condition that is not really grounded in a critical processing of the information they get, but on absorption. This condition is effortless in that it lacks a distanced, lived, temporal perspective. By now, many people do not really get informed by the news they assimilate, they just *in-form* this same news—make them "real" as facts—by simply letting them transfix (para-) consciousness. Already criticizing mass culture, Theodor Adorno (1991, 85) noted that "the curiosity which transforms the world into objects is not objective: it is not concerned with what is known but with the fact of knowing it, with having, with knowledge as a possession. This is precisely how the objects of information are organized today." Various streams of research in cognitive studies support such a standpoint: a technology that is user-friendly (i.e., hides its functioning) and effortlessly provides flows of messages deflates what has been labeled a "generation effect" (van Nimwegen 2008), that is, the cognitive steps required to seek pieces of information to complement the ever-partial picture one is facing. This, in turn, narcotizes people's cognitive faculties: it is well known (cf. Festinger 1957) that people are more alert and willing to engage when confronted with complicated scenarios that literally do not make (straightforward) sense, rather than passively have the whole information stream being delivered to them.

In this sense, while the information spread through mass media created a form of dispossessed, generalized knowledge, today's flow of information does not even allow for that generic knowledge to crystalize. Senders and receivers are entrapped within a logic of mutual deficiency: due to frantic acceleration and overload of information, senders can only guarantee a minimal level of reflection about what they are going to communicate, while receivers can only provide a minimal amount of attention. The meaning of the message, as a possibility to develop a proper discussion around it, is no longer the point; it is all about filling the gaps of daily downtime. And this has deep consequences on how people act (or rather do not act) in the world in that the complete packing of one's own agenda through infodemics triggers a radical withdrawal from actual engagement.

In "The Storyteller," Walter Benjamin (2002) retraced the gradual impoverishment of experience as *Erfahrung*, that is, a collective experience of lived and shared knowledge, replaced by *Erlebnis*, a solipsistic and transient experience. This shift goes along the axis that connects oral storytelling (e.g., the epic) to the birth of the printing press and the diffusion of novels (as textual artifacts that favor solitude and silence to be enjoyed) and to the rise of modern mass media where the practice of sharing knowledge has been reduced to the mere delivery of "facts." Indeed, in a letter to Adorno (Adorno and Benjamin 1999, 140), Benjamin detects a link between the "decline of the aura" enforced by technology and the "coming to an end" of storytelling as a still de-objectified way of recounting and representing the world *in presentia*.

The latest occurrence along this axis is certainly represented by digital technologies, through which socialization has become an algorithmically based form of self-isolation and increased dissemination of news true and false. A few years ago, Stefano Calzati and Roberto Simanowski (2018) conducted a digital ethnography aimed at providing insights into the daily social network diet of young adults. Through this qualitative study, it emerged more clearly that, on social networks, users (1) tend to share materials and reply to others' posts uncritically (without really checking the content of the posts shared or commented on) and (2) forget by and large what they liked/shared after just a few days (being unable to consciously motivate their choices when interrogated).

This leads to a condition in which the public sphere is neither fragmented, as if dispersed into micro public spheres, nor liquid, where individuals would move across different public spheres; rather, the public sphere is sublimated, becoming extremely volatile and ephemeral. Sure, the Web too can be considered a (reworking of the) virtual public sphere in the sense of shared *a*spatial content access. And yet, the difference between the traditional public sphere and the digital one is that the former was based on a dialectics that was possible for the simple fact that a common referent was still there, while the latter configures the juxtaposition and accumulation of pieces of information presented as real for the very reason of being circulated. Hyperexcited by the ongoing swings of emotional waves of news, individuals today live in a very self-centered love affair with information, based on temporary, depthless forms of attachment (indeed a symbiotic dependence) more than on the long-lasting and heuristic acceptance of others' point of view. These others are, by now, beyond recognition.

The Loss of the Signified

According to McLuhan (1964), the "Narcissus Narcosis"—society's fetishist obsession with the contents of literacy—prevented people from recognizing that, by making language visible, it had allowed vision to dominate all the other senses. In the exchange between senses and sense that occurred when writing abstracted meaning from experience, the signified reigned supreme, at least in alphabetic systems (while, as seen, in logographic systems, a symbolization of experience is still engrained in writing). But here comes the season of its demise. The same narcosis is threatening users today as they gaze into their smartphones with the kind of ravishment Narcissus must have felt in front of this fascinating creature (that he did not realize was himself) right there in the pond. The story says that he drowned and became a flower. Another, less bucolic, version suggests that he dried up on the spot, failing to eat or sleep in his total absorption. Dozens of online game players have been found dead in front of their screen for not having taken the time to eat, drink, or sleep for days on end, hypnotized by the game, very much like Narcissus over the pond. Today the myth repeats itself as humans are threatened with extinction by their unchecked adoration for technology. It is about to rid them of meaning altogether. Time has come to highlight the uncanny homonymy of three fundamentals of human behavior: sense, meaning, and direction. These interpretations are expanded on in the *Online Etymology Dictionary*:

> Sense n. c. 1400, "faculty of perception," also "meaning, import, interpretation" (especially of Holy Scripture), from Old French sens "one of the five senses; meaning; wit, understanding" (12c.) and directly from Latin sensus "perception, feeling, undertaking, meaning," from sentire "perceive, feel, know," probably a figurative use of a literally meaning "to find one's way," or "to go mentally," from PIE root *sent- "to go" (source also of Old High German sinnan "to go, travel, strive after, have in mind, perceive," German Sinn "sense, mind," Old English sið "way, journey," Old Irish set, Welsh hynt "way"). Application to any one of the external or outward senses (touch, sight, hearing, etc.) in English, first recorded in the 1520s.

Interesting to note in the above, the connotative values of the radical *sens* from the sensory to the mental realms, including direction, are in all the term's appearances, at any epoch, starting with Latin. Meaning is the dominant meta-sense that brings the other two under control. For humans and, one can assume, many other living beings, senses and utterances are at the service of meaning to guide action or direction. But, as algorithms are taking over the function of control, there is no need for meaning at all. Are the senses obsolete? 0/1, the lowest common denominator of all the digital translations of sensory inputs, re-creates general synesthesia via multimedia and virtual /reality/. It is very likely that digitally enhanced senses will lead to the atrophy of the biological ones, downgrading them from survival to entertainment value.

"I wouldn't have seen it, if I hadn't believed it," McLuhan quipped. Human eyes perceive only a small part of the electromagnetic spectrum, wavelengths between 400 and 700 nanometers, but humans cannot perceive ultraviolet, infrared, gamma, and X-ray rays. The same applies to human hearing, which can only perceive frequencies in the Hertzian acoustic range from 20 to circa 20,000 cycles. So, to overcome the limits of biological senses, and also extend their reach to the whole of the environment, technology provides and distributes new avenues for sensing: "Sensors are used to monitor climate change, human settlement processes, the behavior of animal species and the relationship between all these aspects. In the end, therefore, we witness a sensory revolution of nature itself" (Accoto 2017, 43).

A sensory revolution of the body itself is indeed on course with the foreseeable industry of implants on the horizon. Take the example of the recent invention of a "biomagnetic sensor." With this magnet, it is possible to overcome the limits of the biological hardware. An expert biohacker added a sixth sense: the ability to sense magnetic fields. It is a unique—and one could say extreme—case of someone who is driven by the desire to try a technologically enhanced sensory system:

> In recent years, the empire of sensors and senses has been gradually expanded to cover the whole world: bodies and objects, even spaces and environments. . . . From micro to macro scale, we are building a new sensory system. The sensors and data that we are incorporating into the world redefine our concept of environment while at the same time being part of both the new home ecology and the non-dominated environment. (Accoto 2017, 41)

The third major meaning of sense has to do with direction. It is the only one remaining in algorithms. To recap, according to Vico (1744), the survival of humanity began with the senses, the first algorithms of all mammals. With the development of language, the algorithmic power of the senses was amplified. Writing further reduced the sensory domain by provoking an exchange between senses and sense and giving priority to language. Writing took over the predominant algorithmic function from oral language and reduced the senses to its (visual) service, and this led to the extremes of representationalism that favors words over matter. Today, the faster, more effective digital algorithm of command and control has eliminated the use of both senses and meaning. That is one side, the more individualistic one, of the digital transformation. The other side is how it also restructures society. Perhaps the social direction (sense) is that social control is evolving from bureaucracy to data power. Through such evolution, the embodiment of sense allowed by language is further obsolesced:

bureaucracy attaches meaning to objectified entities, but data objectify meaning-making itself; the direction is only one of accountability and performativity. As Guattari (1995) astutely remarks in his *Chaosmosis*, one should either suspend any logocentric understanding of "reality" or at least couple it with a machine-centric one to reconfigure the functions of language from a meaning-producing technology to a sign-based world-sensing that is presubjective.

Trusting the Algorithm

The fuel of datafication (the translation of life and social processes into sets of data) is, quite evidently, big data, namely, the aggregation of large data sets that can be triangulated in order to be analyzed and eventually reach decisions (sometimes automated). Significantly, a body of academic works has revealed the extent to which such datafication is not neutral—no translation is—but has deep sociocultural and ethical implications (not least because many actors providing and harvesting data-driven solutions are private corporations). Moving past these relevant contributions and pointing directly to the philosophical core of the issue at stake, one provocative question would be, "Can one get rid of data and analyze and decide *without* data?" No doubt, taken as an abstract generalization, this question may sound paradoxical or even absurd, and yet, once scrutinized, it invites one to explore what one means by data.

Reconsidering Data (and Information)

Following a normative understanding, data are considered the unrefined version of information. Silver and Silver (1989, 6), for instance, define data as "the raw material that is processed and refined to generate information." This is a realist stance (i.e., one that regards data as true and factual by default) that leads to consider information as the meaningful organization of data, assuming that the two are inherently of the same kind and the passage from one to the other can occur linearly and smoothly. Such a stance defines, de facto, how information and data are understood within information science, with the data–information–knowledge–wisdom (DIKW) pyramid as a trademark. As Martin Frické (2009) writes, scholars subscribing to this pyramid consider the difference between information and data as functional, not structural.

However, it is possible to not only subvert such relations—with information preceding data—but argue in favor of the nonisomorphism between the two. This allows us to deconstruct the "naturalization" of data as a resource and to emphasize, by contrast, their sociotechnical fabric. According to information theory, information is fundamentally constitutive of reality, to the point that physicist Wheeler summarized this idea with the slogan "it from bit." In a similar fashion, Gregory Bateson (2000, 21) nails this down, affirming that "information is difference that makes a difference," meaning that, beyond metaphorical and symbolic communications and even beyond the processing of digits that appears as increasingly autonomous, at its core, information is the recording of change, a differential process.

From here, it follows that data are always already the bio-sociotechnical seizing of certain information. As Rolf Landauer (1996, 188) puts it nicely, "Information is not a disembodied abstract entity; it is always tied to a physical representation." Put differently, information is syntactic in principle but can only be grasped semantically. On this point, the discussion by Zellini (2022, 328) on continuum and discrete comes in handy. "Even assuming," he notes, "that the continuum exists in fact, we would only be able to describe it through a gradual fulfillment realized via refined techniques of approximation in the discrete." In other words, the continuum can only be thought of as an analytical concept, an act of cognitive representation; however, "even if unknowable, the continuum remains an unavoidable presupposition" to make sense of the discrete (i.e., of physical reality). Paralleling this argument, one might think of information as an unmarked field that becomes pertinent only once it is materialized under certain conditions: "Our world is pregnant with information. It is not an amorphous soup of atoms," César Hidalgo (2015, xix) writes, "but a neatly organized collection of structures [that] are the manifestation of information." Data do not preexist in nature; instead, they are a human and/or tech-created construct—a *certain* enframing of /reality/—that exists in the very moment in which a *certain* process is enacted to collect *certain* information. Translating the continuum into data, rather than just an invention to create a more homogeneous, fluid, and manageable representation of things than words or texts, may have been a necessary step of evolution to arrive at a quantum level of resolution.

Overall, the grasping of information always occurs through embodiment as a bio-sociotechnical disposition that is at once mechanical (e.g., tools), organic (e.g., living bodies, cognitive faculties), and ideational (e.g., means of expressions, values). While distinct for the sake of the argument, these aspects are coexistent. When it comes to mechanical means, Charlotte Hess and Elinor Ostrom (2007) explain that "this ability [of technology] to capture the previously uncapturable creates a fundamental change in the nature of the resource, with the resource being converted from a nonrivalrous, nonexclusive public good into a common-pool resource that needs to be managed, monitored, and protected, to ensure sustainability and preservation." An example comes from the recently detected gravitational waves (or also radio waves detected by Heinrich Hertz at the end of nineteenth century): it is not that these waves were inexistent before—in fact, they had been predicted—but they were unmarked information, which then turned into structured data when a certain technology and measurement became possible. Concerning value, just think of any occurrence in which people mobilize their interests toward a given un(re)marked physical feature of the environment—be it a mineral component or the configuration of landscape. As they enframe it through the lens of their shared value—say, for energy or entertainment—they instantly metamorphose such, until then, irrelevant feature into a pertinent resource. Keller Easterling (2021, 62) wisely notes that "it is the architecture of interplay and entanglement that is the real innovation. . . . Value begins with physical arrangement, location, community, diversity." Last, concerning the organic dimension, the grasping of information is dictated by and through focal attention—or consciousness (cf. chapter 5). The open-ended fact of simply living, intended as an immanent and non-predetermined process, repeatedly finetunes with the environment and orients, summons, and ultimately actualizes certain physical instantiations, based on certain psychophysical needs, and in view of certain goals. To live is, first of all, a matter of ecological dispositions, of "search and see" (Gibson 1966). Most important, such disposition, dictated by the triad agent-goal-instantiation, emerges as a unique entangled whole (Walsh 2015).

Through digitalization, it is not only that an increasing number of phenomena get computed but also that this process is transforming how people value the world, which means how they live and who they are. This is an eminently qualitative issue that implicitly asks what, through digits, one *can* know and *how*, as well as what one *cannot* know and *why*.

The Epistemology of Digitalization

Unfortunately, when it comes to digitalization, the answer often provided to the issue above relies on a quantitative rationale: "With enough data, the numbers speak for themselves," Chris Anderson (2008) notoriously wrote preconizing the end of theory. Yet what is often overlooked is the epistemological nature of data: while it might already be admitted that data are not neutral or raw, insofar as they embed precise cultural values (data as agencies), they also have a performative side (i.e., they are agents) and, as such, they (re) enact a precise world-sensing. This world-sensing finds its roots in underlying positivist approaches to what people assume to be real. This means that data, as bits that can be collected and shared, inform an understanding of analysis and decision as (already) objective/objectified processes. These processes, indeed, are based upon a condition *sine qua non*, notably accountability. To "ac-count" draws originally upon the idea of describing by counting, which inevitably means, as Werner Sombart (1987, 119) argues, "to pursue the basic thought to grasp all phenomena only in quantities, the basic thought, thus, of quantification." This claim brings to the surface the systemic unavoidable partiality of all data—in the sense of being parcels of /reality/—and consequently, it questions the soundness of the whole analysis/decision enforced, as it emerges, for instance, when data-driven technologies are misled to over-correlate statistical relationships producing incorrect or undesired results (ChatGPT seems to establish a legitimate order of information, which is quite coarse at the moment, even though it might soon raise to an agreed referent). "The fundamental problem," Wolfgang Drechsler (2019, 230) contends, "is that if [data] necessarily show only partial aspects . . . then this means that one can always construct a set of indicators that proves any answer one wants to the question posed." From here, we arrive at the core issue. Analytical and decisional processes driven by data are considered reliable for the very fact that data are accountable. A fetishization of data, assumed to be the best lens through which to interpret the world, is at stake but hardly recognized, trapped in a sort of tautological spell and suspension of distrust. It is not only that people are datafying (and automating) an increasing number of processes but that this datafication is getting prominence over what has not been (cannot be) datafied yet. In a social context drowning in fake news and denial, the danger is that either the assumed distinction between objective and subjective gets lost or datafication becomes the one and only locus of objectivity.

To be sure, at stake here is not an apology of irrational thinking, but the very possibility that there are phenomena falling beyond the immediate rationalization and positivist understanding that data bring with themselves or, alternatively, that there are other means of signification to grasp and mold reality and give sense to it. Extending Gödel's theorem to the physical realm, Devereaux and colleagues (2021) argue that no logical modeling of the world can ever be achieved. In *Gödel, Escher, Bach*, Hofstadter (1979, 395) acutely notes, speaking of computation, that "above a certain level of complexity a qualitative difference appears, so that 'super-critical' machines will be quite unlike the simple ones hitherto envisaged." Hofstadter goes on to write that

> we can liken real-world thought processes to a tree whose visible part stands sturdily above ground but depends vitally on its invisible roots. In this case the roots symbolize complex processes which take place below the conscious level whose effects permeate the way we think but of which we are unaware. . . . Real-world thinking is quite different from what happens when we do a multiplication of two numbers, where everything is "above ground," so to speak. . . . When it comes to real-world understanding, it seems that there is no simple way to skim off the top level, and program it alone. The triggering patterns of symbols are just too complex. . . . The point is simply that meaning can exist on two or more different levels of a symbol-handling system, and along with meaning, rightness and wrongness can exist on all those levels. (563)

This excerpt is rich in implications. By comparing computers and the brain with respect to how they process the world, Hofstadter suggests that there is a threshold above which mere computation cannot do the whole job alone. The whole is irreducible to a breaking down into simpler (algorithmic) processes. Information and data are two incommensurable systems. The hiatus that Hofstadter identifies is between a way of abstract (logical, mathematical) thinking and a real-world (computational, statistical) one. This hiatus is qualitative, not quantitative, so that even if/when the aggregation of huge sets of data will eventually "compensate" for phenomena that cannot be computed, what will come up is not simply the completion of the puzzle; rather, an entirely (epistemologically) different dimension will be fostered. And in such a dimension of "hyperdatafication," the very act of discerning (analyzing and deciding) will totally escape anthropological understanding.

The main claim here is that the accumulation and aggregation of big data—thickening the data-driven shaping of /reality/ and delegating to data an increasing power for analysis and decision—is weakening the (language-based) epistemological pillars of society and, especially, that same rationalist-positivist view that has led to today's datafication in the first place. By entering in this new dimension in which analysis and decision are heavily dependent upon automated processes escaping scrutiny, values such as truth and falseness, good and bad, conflate and are copresent, at least from an anthropological perspective. But then, if (rational) discernment escapes examination, will decisions be *really* good? For *whom*? "If the citizen loses trust in the state's knowledge and thus in the rationality of its decisions, the readiness to follow sovereignty commands evaporates," Andreas Voßkuhle (2008, 18) warns. According to him, the possibility for citizens to monitor the state's rational, analytical, and decisional processes— that is, to keep the state accountable for its actions—is what incites citizens to comply with the state's sovereignty. The more such accountability fades, the more citizens will question the state sovereignty. It is hardly surprising, then, that today, with an increasing number of processes being automated (including the digitalization of traditional bureaucratic practices), there is a widespread erosion of citizens' trust in institutions and authority.

To be sure, data and data-driven processes are not obscure per se (the usual black box metaphor); it is rather what surrounds them that is opaque, politically and epistemologically. The disappearance of the referent, as examined above, hits the very possibility of sharing a common ground on which to build any pretension to truth. In a society increasingly guided— if not yet dominated—by data-driven technologies, even such pretension is not in question, but the idea of truth itself. In fact, the discrimination between truth and falseness becomes a matter of probability, more than certainty; the poles of objectivity and subjectivity conflate.

The most epitomizing case is "deep fakes," the phenomenon described above that better than any other shows, beyond the technical power of digital fabrication, the superposition of traditionally distinct epistemological states. Deep fakes are real *and* false at the same time: of course, their disentanglement is (still) possible (with a good dose of criticism), but this is beyond contention. Deep fakes, indeed, bypass reason by leveraging on (the fascination of) being "paradoxical," just another possible opinion, which is appealing for the very fact of destabilizing the known in favor of the probable. This is why the term *post-truth* is somewhat misleading: society has not moved beyond truth, nor has truth disappeared; it is still around, but it has become probabilistic, extremely conditioned, and forced to accept to be merged with its own opposite.

Ecosystemic Knowledge

To the extent digitalization is one (quantitative) way of framing information, there will always exist other (qualitative) ways of doing that. Generally speaking, these ways are *practices*, considered as *modi operandi*, that is, (embodied) *ways of doing* that inevitably depend on socially shared agreements. Practices defined as such get close to what is sometimes referred to in information science as "know-how" in opposition to "know-that" (which identifies declarative statements on states of the world). If here we prefer to speak of practices, it is because the distinction between know-how and know-that overlooks the *dispositif*-dependent codification of data, statements, and practices: literally, "we know more than we can tell" (Polanyi [1966] 2009), but we also express more than we know, or better, knowing is a matter of ever-different *dispositif*-dependent configurations. To claim, for instance, that know-how, differently from know-that, cannot be recorded (Fricke 2009) because it is inexpressible comes from imagining data and/ or language as the basis of some superior form of evidence that can be objectively transmitted. A key point to which not enough weight is given is that language is never "natural," as much as data are never "raw" (nor is the body, for that matter). At stake is a critique to foundational logic through the questioning of the materiality of any expressive form. In fact, it *is* possible to convey know-how through forms of social (collective) practices, for instance, by building upon expressive forms other than data and language (including artforms and performances that co-opt the body as a technology). We might even say that declarative statements in the form of know-that are linguistic approximations of know-how as embodied (tacit/explicit) practices, in the same way that digital models deliver a computable formalization of /reality/ approximating complex physical phenomena. This is also why truth and factuality, which information science tends to attach to data and propositional statements by default, are not ontological properties of either data or statements but sociohistorical and collectively defined values.

This resonates with Benjamin's distinction introduced above between *Erfahrung* and *Erlebnis*, which now appears under a new light. While concretizing a techno-based experience of the world as *Erlebnis*, data do originate from socially shared practices as *Erfahrung*. This is where Benjamin's stand-point betrays a certain longing for origins (speech), which tends to over-look the embodiment of any communicative act and knowledge. In other words, one can never have a pure form of *Erfahrung* as opposed to *Erlebnis*; rather, it is important to explore how a certain disposition transforms infor-mation along the axis connecting the two realms. A crucial example is law, as a form of reification of "reality" through language: "a patent applicant," Brett Frischmann and colleagues (2014, 23) write concerning intellectual property rights, "must demonstrate that the invention claimed in the application possesses an 'inventive step,' such that the invention represents a suffi-ciently great technical advance over the existing art." This epitomizes how law, by means of language, dissects experience (as *Erfahrung*) and turns it into *Erlebnisse* (ready to be economized). Law artificially creates rights (value) by parceling human activity (in the same way as technology turns a public good into a limited resource). At the core of both lies the schism between object and subject—*res cogitans* and *res extensa*—that an ecological approach can help undo and rework (*ego-res mutans*).

Then, between data and practices a dialectic space exists: in this regard, knowledge is, yes, connected to data and practices and yet stands on a dif-ferent level; it is a meta-reflection on information—the distillation of all that people can synthesize from data and practices as an *organized process of signification*. Far from collapsing into information (Frické 2009), knowl-edge is a *meaningful arrangement of data and practices*, that is, a form of self-organization that fundamentally (and endlessly) seeks, processes, and passes on information. According to Chiara Merletto (2022), physical reality is governed by "laws without a project" (without a finalist "intention") and yet they are laws that create room for what is possible and what is not pos-sible. In nature, there are no goals or projects in the traditional sense of the term: what is there is a fundamental dynamic of possibility(ies). One might see this dynamic as information in constant formation: indeed, Merletto defines knowledge as "resilient information," that is, information (about the possible) that "resists" and can be passed on over time. This prefigures an ecosystemic understanding of knowledge that implies that the point of contention is not who owns knowledge or where it is supposedly located but how a certain knowledge is bio-sociotechnically embodied and trans-mitted: "The fact of living . . . is to know in the realm of existence" (Mat-urana and Varela 1987, 174). Since one is born, "it" embodies a certain

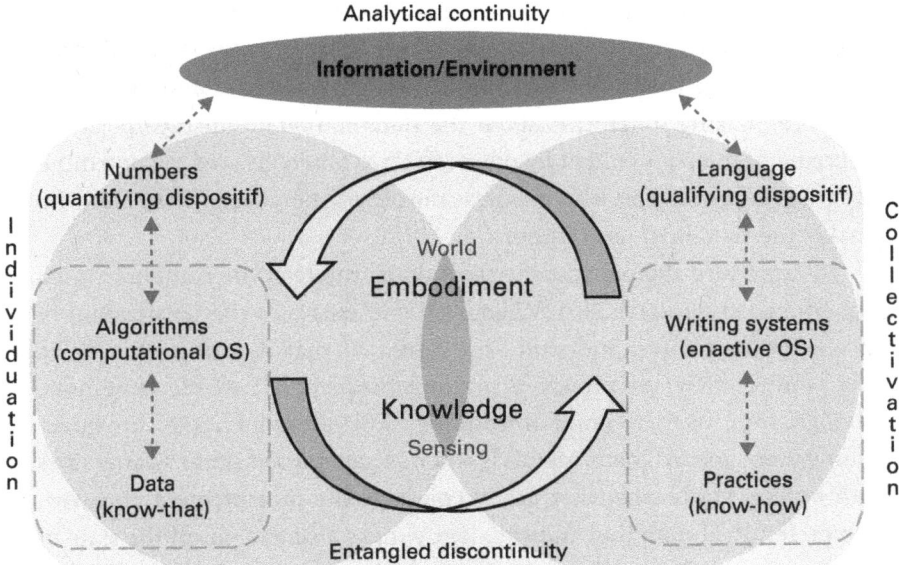

Figure 2.2
Framework for the data–information–knowledge ecology.

world-sensing through bio-psycho-physical-sociotechnical dispositions. This leads to updating the figure provided in the Introduction about the language and digital ecologies, with a comprehensive overview of an ecological understanding of knowledge that includes data and practices (figure 2.2), whereby data are associated with declarative know-that (closer to the ethos of *Erlebnis*) and practices are associated with language (closer to the ethos of *Erfahrung*), with embodiment always mediating between these *dispositifs*.

*   *   *

At the level of quantum physics, an extremely interesting experiment in support of the copresence of complementary epistemological frameworks (i.e., loss of the referent as per our discussion) has been conducted recently at Heriot-Watt University in Edinburgh. A team of researchers (Proietti et al. 2019) performed in the laboratory the critical Wigner's thought experiment. This experiment, elaborated in 1961 by physicist and Nobel Prize winner Eugene Wigner, deals with the indirect observation of a quantum measurement, creating an apparent paradox between two competing views on the measurement of a quantum system, which in turn questions the existence of an objective reality to which all subjects can refer. In fact, the experiment aims at proving that when two observers—say, Wigner and a friend—relate to the "same" event from different (nestled) standpoints—for instance, making statements about the state of a quantum superposition, subsequent to measurement by one of the two observers—an incompatibility arises depending on who performs the observation (and when) and who makes the statement (and when).

To reproduce the experiment conditions, the team in Edinburgh used six entangled photons. First, Wigner's friend measures the polarization of a superposed photon and stores the result. At that moment, then, from her point of view, the photon is no longer superposed. At the same time, though, from Wigner's point of view, the whole experiment (i.e., the quantum system and Wigner's friend) is still in a superposed state: Wigner does not exactly know when her friend conducts the measurement and with which result. Should they independently make a statement on the state of the quantum system, these would be contradictory. Then, Wigner performs his own measurement, or alternatively, he asks her friend about the result of her measurement: from Wigner's point of view, it is only at this time that the superposition disappears, and the photon takes a definite state, even though, from her friend's point of view, the superposition had already disappeared. The conducted experiment gave the result that Wigner predicted, notably that the two alternative realities of Wigner and his friend could coexist *even though* they were actually based on irreconcilable standpoints.

This suggests that there is not one privileged observer position and a unique ur-referent to which to relate; at stake are always complementary scenarios (or "local observer independence" from the title of the article describing the realization of the experiment). Most subtly, this has deep epistemological consequences. The contexts affect what people think and what they (can) know. This is a sort of Copernican revolution of epistemology, coming right from within science. One gloomy interpretation might be a sort of self-declared solipsistic position: something like "I only know what I can know from where I am and who I am." The more proactive stance would be, however, "as soon as I 'move,' a new reality opens." In fact, living can be regarded as the enactment of a *dispositif*-dependent information asymmetry repeatedly reconfigured with regard to both the outside—the environment—and the inside—one's consciousness—of which embodiment constitutes the barycenter. Such asymmetry can never be solved; it is open: in fact, the more one knows, the more one ignores.

Just as the formalization of the point of view became the condition for individual freedom in the neutral space of Renaissant perspectivism, a proprioceptive appreciation for one's point-of-being in the networked data flow is among the conditions for retaining a measure of physiological and psychological control over one's whereabouts in electronic nomadism. The idea of a point-of-being enables one to keep track of oneself when the technologically extended senses are operating all over the planet. The physical sensation of being somewhere is environmental, not frontal. It is comprehensive, not exclusive. One's point-of-being, instead of distancing one from the out-there given, like a point of view, becomes a point of entry into sharing the world and rethinking the triadic relation of human–environment–technology. Now, the combination of point of view and point-of-being metamorphoses into a field of probability—of both the sentient beings and the environment with which they are entangled—which can only be iteratively disentangled through embodiment.

Self-Sabotage and the Return of the Absurd

Self-sabotaging behaviors and absurdity are two major symptoms of the increasing difficulty people have to make sense of the emerging digital ecology. These two symptoms are very much intertwined, to the point that they are cause and effect of each other.

Self-Sabotaging Behaviors

Sabotage comes from *sabots*, which are wooden shoes. According to socialist John Spargo (1913), writing at the beginning of twentieth century, "In France it has long been the custom to liken the slow and clumsy worker to one wearing wooden shoes, called 'sabots.'" From here, *sabotage* comes to define a deliberately clumsy move by workers, to surreptitiously hijack production in factories. *Saboter*, then, originally meant to obfuscate, derailing the normal flow of the human–machinery cooperation, a more effective alternative to strikes or political mobilization because it disrupted from the inside the functioning of the system. In other words, the *saboteurs* were people who tried to regain the upper hand on their employers through secretly alienating the machine. According to psychologist and scholar Michael Vannoy Adams (2012), a Freudian reading of sabotaging traits can be understood as "how the ego represses—or, more specifically, sabotages— what the unconscious attempts to express." The idea that under the drive to sabotage there is a conflict between conscious and unconscious, deeply resonates with what we have discussed above, notably that technology has forcefully exteriorized onto the social stage the individual unconscious and disjoined it from any relevant, or even understandable, meaningfulness. Vannoy Adams continues,

> From under the ego, the unconscious attempts to overturn the ego. From the perspective of the ego, the unconscious is intrinsically subversive. No wonder the ego is so anxious and so defensive. Ultimately, the unconscious as such is a saboteur, for it attempts to sabotage—or to subvert—the partial, prejudicial attitude of the ego.

Thus, as soon as the (digital) unconscious, intended as an agglomeration of data that informs data-driven processes of which people are not fully aware, goes public, this externalization conflicts with a shared understanding of the subject and "reality." Here the term *conflict* is crucial: once the common ground is gone, this conflict comes to affect both the individual and the collective, against their own interests. And while it might be easy to fear and fight the Machine as a Moloch (represented by Fritz Lang in *Metropolis*), it is much more difficult to detach oneself from commoditized technology turned playful. At worst, users are willing to subject themselves to technology; at best, even when they do want to hijack technology, they are drawn into a mirrors' room, finding themselves shooting eventually at idols, rather than at the true (data) centers of power.

One example of individual self-sabotage is selfies, a practice that is "social" to the extent it nurtures the connected virtuality of the Web, while increasingly eroding the uniqueness of the subject through being constantly reproduced. On a closer look, there is nothing narcissistic about selfies, insofar as the image they convey is less and less self-sufficient, the more it gets distributed. People show off online in order to accumulate social acceptance and get the illusion they belong to a whole, at least for the time of a dopamine-producing click. After all, if people's unconscious has been exteriorized in the form of a commodity (likes, sharing, etc.), which everybody craves for, the conscious on the social stage cannot but bear the traces of such anxiety in the form of a many-to-many competition. By formatting /reality/ and human experience into binary code, digitalization has reduced the richness of language-based technocultural fields, leaving people root-less and prone to clash. And this, in turn, contributes to further reinforce a paradigm that makes networking and globalization its main traits: exterior-ization and transparency of peer-to-peer communication—especially when guided by algorithmic juxtaposition that creates increasingly self-sealed fil-ter bubbles—produce a conflicting society where gray grounds for mutual understanding are canceled by the very binary logic of digitalization.

Notwithstanding the ephemeral inconsistency of global political deci-sions, led more by transient sentiments than long-term evaluations, which then evaporate in a total lack of vision before major threats such as climate change or pandemics, one example of collective self-defensive self-sabotage is nationalism (precisely because inscribed into a global perspective as a sort of superego ruling on the ego). Nationalisms are the symptoms of an anxiety that is not external but endemic to technocultural fields unable to cope with the formatting logic of global interconnectedness. London-based writer and consultant Umair Haque sums it all on his website dedicated to *eudaimonics*—the life well spent and worth living—as follows with regard to Brexit: "Brits have become people incapable of thinking sensibly in the most basic or fundamental ways."[7] What he does not sufficiently recog-nize is that the roots of such behavior are not political, or educational, but psycho-technological. It remains to see to what extent such drives will also come to affect Asian countries, especially China, once the system has exhausted its socioeconomic force: (nationalistic) signs, in conjunction with deflation and unemployment, are already there. Yet, in this scenario, the peculiar entanglement between logographic language and collective culture might well produce an idiosyncratic result.

The Return of the Absurd

On May 23, 2019, a picture taken by expert Nepalese alpinist Nirmal Pujra was relaunched by all the major broadcasters and newspapers worldwide and went rapidly viral on the Net. The picture shows an astonishing long queue of climbers trying to reach the summit of Mount Everest. Climbers are in life-threatening conditions (cold, limited oxygen, exhaustion) and, despite that, they are caught waiting in line to slowly walk the last few hun-dred meters. For expeditions aiming for the highest peak on Earth, 2019 was a particularly bad year: a record number of climbers facing a particu-larly unfavorable climbing season, with only a few days of good weather during which to attempt the ascent. It was these conditions that led to the formation of the long queue photographed by Pujra, and most unfortunately, eleven deaths were reported during those days. Filmmaker Elia Sailaky summed up the situation as "Death. Carnage. Chaos. Line-ups. Dead bodies on the route."[8] Pujra's picture has a subtle value: it depicts well the heights of absurdity reached by today's society; it witnesses the extent to which the politics of selfies—the nihilistic self-absorption it demands and promotes—has reached out onto "real life"; it stands for the filter bubble in "real space," and it tells a lot about the power of self-destruction implicit in the tendency to turn today's life—even its most serious aspects—into a village fair competing carrousel. The issue is not having taken the risk to die on a mountain's ascent; it is to die, unaware of one's obsessive egocentric craving for self-exposure. This is symptomatic of the resurgence of the absurd, not so much as a feeling but as an epochal ethos.

*   *   *

The Notre Dame fire in Paris, also back in 2019, could perhaps be interpreted as a sort of "quantum effect." I (de Kerckhove) remember talking about absurdism with McLuhan, who never failed to reconstruct etymologies at will and occasionally at whim. When I suggested (wrongly) that *absurd* was probably derived from *ab* (without) and *surd* (root), he was thrilled because he was trying to explain why he thought that in some specific cases, "effects could precede causes," notably with great artwork that he felt had a predictive rather than merely a reporting value.[9] To my suggestion, he said, "Yes, that's it, rootless, without ground," a reference to his pet theory of the Gestalt relationships between figure and ground. Unfortunately, the etymological dictionaries all agree that *surdus* means "deaf," turning the etymology around into "from a deaf ear," by extension, from someone who is deaf to reason, who knows nothing. But for the sake of discussion, let's keep *reductio ad absurdum* as my original interpretation: what is absurd is groundless. So is quantum, which indeed builds upon an *écart* (the Planck's constant) as not only a gap but also a limen of possibility. The useful insight here is that quantum is really an effect, not a cause. The most precise definition is the one we have chosen, quantum ecology. Being self-organizing, ecology, like quanta, has effects, but not causes, at least not causes in the ordinary sense of a direct relationship between cause and effect. This resonates with Lee Smolin's idea about the possibility of nature "developing habits as it goes along," leading to the (heterodox) suggestion that "maybe if you make a really novel system then we'll have no precedence—it won't behave as you expect it to because it won't know what to do, so it will just give you some totally random outcome."[10]

Four years before her celebrated *Meeting the Universe Half-Way* (2007), in a footnote to a paper that could serve as a preface to it, Barad (2003, 814) makes an important point about Heisenberg's principle: "The so-called uncertainty principle in quantum physics is not a matter of 'uncertainty' at all but rather of indeterminacy." We have sensed the need for that distinction almost from the beginning and tried to elucidate it with a progression of "steps to probability":

1. Uncertainty/doubt: the focus is on the observer
2. Indeterminacy: the focus is on the observed
3. Probability: the focus is on the outcome
4. Measurement: a determination of the outcome
5. Causality: an estimation of cause–effect relation (only after the measurement)

Interestingly, even as she refutes any possibility of it as a preexisting condition, Barad comes back again and again to save causality on her "agential-realist" terms. For example: "the agential cut enacts a local causal structure among 'components' of a phenomenon in the marking of the 'measuring agencies' ('effect') by the 'measured object' ('cause'). Hence, *the notion of intra-actions constitutes a reworking of the traditional notion of causality*" (2003, 815). According to Barad, mattering—the coming into being of reality— is produced as an ongoing immanent ever-implicated "cut" within reality itself. Bearing this approach upon the surge of the fire in the cathedral, we can propose an uncanny "effects before cause" phenomenon.

Like quantum effects, ecologies emerge as entanglements of a plurality of relationships. With respect to coherence and decoherence, one can imagine that, for example, quanta morph (as in morphogenesis) into shapes by the phenomenon of "strange attractors." The absolute novelty of the quantum is that the effects cross all boundaries: everything is in constant implication. The quanta traverse the physical and the cognitive with equal ease and modify both. They call for cognitive abduction as much as they impose sensible prehension. And, because quanta also deal with immaterial occurrences such as ideas and thinking or emotions, and perhaps even dreams, we suspect that this can bypass boundaries such as standard logic (cf. Priest 2007). For example—and we confess right away that this is absurd—the burning of Notre Dame cathedral in Paris could be the quantum effect of the concentration of global anxiety, expressed by the global attention paid to the event and sustained semiconsciously by its enormous symbolic meaning. All those faces evidencing fear and dismay around the burning site were ready long before the event was to produce this emotion.

We are working on precisely what effects would precede causes or, better, be potentially copresent. The case of the cathedral is a good hypothesis to start with because the event itself is first highly unlikely, considering the inherent structural and material stability of cathedrals (it takes modern wars to really bring them down, as in Cologne, for instance). Second, it occurs at a genuine revolutionary time in the world in general. Everybody is nervous about everything now, in a way that is unprecedented since the early tidings of World War II. The cognitive dimension of the quanta could be supposed to trigger the physical occurrence as a self-organizing response to a genuine human trauma, underlining its vast symbolic connotations at architectural, social, religious, temporal, and nationalistic values, to mention just a few. All this does not confer a "causal" role to quanta, only a physical one, as the way by which a morphogenetic concurrence and confluence of human emotional conditions brings about a *catastrophe* in René Thom's acceptance of the word, that is, something that occurs only after a certain level of accumulation of various elements defines the moment of a rapid and brusque change from stability to breakdown.

Looking Ahead

Where does one store fears and pities? People construct shields in their bodies as revealed by Gestalt and stress therapies. And as the social body achieves a global outreach—however torn it might be—it needs new self-defensive strategies. What would *catharsis* be like at a global scale?

Catharsis is (mis)understood generally as "purification." This interpretation relates it well to *miasma*, which for the ancient Greeks was a form of pollution, a corrupted social situation. Other approaches point to relief, release of psychological tension, but few suggest the deeper meaning as "understanding." According to Aristotle, catharsis arises from *anagnorisis*, which means recognition. The global equivalent to today's needed *anagnorisis* is to recognize that today's upheavals are momentous manifestations of a deeper trend, that of the self-organizing process of humanity trying to restructure itself from/through language and digital ecologies. It is indeed violent and uncomfortable, but once the fact is known, new behaviors can be imagined, learned, and practiced.

In the next chapters, we will look into this: how the digital transformation is creating the preconditions for a new (quantum-based) sociotechnical paradigm and how we might look for a new (ecological) order.

3 Digitalization: Entangled Datacracy

The metaverse is a persistent and immersive simulated world that is experienced in the first person by large groups of simultaneous users who share a strong sense of mutual presence. It can be fully virtual, or it can be layers of virtual content overlaid on the real world.
—Louis Rosenberg (2022)

Today's metaverse is just the peak of the (tech) iceberg, a peak that is melting quickly and will soon turn into an ocean of disruptive changes. The metaverse and whatever is to follow constitute the condition *sine qua non* for a total environment with instant real time to and from data and communication, fostering a sort of tech–human–context symbiosis.

We call datacracy—the power of data—the ultimate sociopolitical dimension of the digital transformation. Datacracy configures a de-territorialized, de-subjectified, time-void condition whereby the consequentiality of action collapses in time and space up to being engineered for automated predictability. In fact, such predictability emerges from the recombination of data by algorithms, not human thinking; the rationale of datacracy is performative overcorrelation, falling well beyond human understanding; the subject becomes a transfixed data agglomeration. It is the formatting operating system that plays and arranges its (f)actors (and discourses), not the other way around. The key is in the dis/trust people (think they) have in the system, although such dis/trust may dissolve into blind abandonment. Datacracy is a ruling anonymous force that subjects individuals to coarse accountability. And yet, datacracy's own accountability—political and technical at once—is increasingly out of reach. To be sure, the power of data and the execution of such power do not configure an all-encompassing panopticon system.[1] Nonetheless, the digital *dispositif* is edging toward an invisible smoothening and seamless refinement of such power and execution.

In the first part of the chapter, we will outline the main (f)actors and features of datacracy, while in the second part, we will discuss its main applications and discourses.

Data(field)

Datafication works as a leveling force that translates the multifarious diversity of physical reality in terms of binary code. Such translation then makes the engineering of /reality/ possible because its facets and components have been formalized in such a way that they can all be integrated as interoperable interlinking gears. The loss of specificity (and meaningfulness) is compensated by the gain in recombination. One thing that is increasingly evident today is that globalization is an issue not just of good or bad effects of practices, policies, and measures affecting the whole world but rather of entangled fate. Such entanglement is based not so much on connectivity principles (i.e., binding links between two contexts) but on instant reverberation principles depending on hypercorrelation.

As soon as this global codependency is formalized as data, a whole new dimension, which is also an arena of action, is being created. We call this arena *datafield*. This notion draws upon Arjun Appadurai's (1990) idea of global cultural flows. Appadurai characterizes these flows as *scapes*, a word meant to define "not objectively given relations . . . but rather deeply perspectival constructs" (1990, 296). He also claims that the concept of *scape* allows "to evoke certain technologies or institutions without confining them to a single location. It is trying to identify some basic links between the conditions of material life and the conditions of art and imagination" (in Rantanen 2006, 7). The term *scape*, then, defines a certain conformation of the economical, political, and social order. Appadurai's identification of fives "scapes"—ethnoscapes, technoscapes, financescapes, mediascapes, ideascapes—aims at accounting for the ways in which this same order is constantly reshaped at various points of juncture and disjuncture.

Of particular interest here are the notions of "technoscape" and "mediascape." By the former term, Appadurai (in Rantanen 2006, 15) means "the global configuration of technology, and of the fact that technology, both high and low, both mechanical and informational, now moves at high speeds across various kinds of previously impervious boundaries"; by the latter term, he refers "both to the distribution of the electronic capabilities to produce and disseminate information (newspapers, magazines, television stations, film production studios, etc.) . . . and the images of the world created by these media." Hence, techno and mediascapes materialize at the intersections of channels, contents, and discourses. Appadurai also specifies that "[the internet] would be in both, as a matter of interest" (in Rantanen 2006, 16). This brings us back to chapter 2, creating a parallel with the observation that algorithms are a special kind of operating system, one that, albeit creating a certain configuration of the world, does not have any meaningful message to be communicated.

Unfortunately, the connotations of *scape*, as Appadurai's examples suggest, retain a certain fixity and visuality; by contrast, starting with the notion and imagination of *field*, it becomes easier to understand how datafication emerges as an increasingly entangled phenomenon. Being the object of a point of view, *scape* is exclusive of the subject/observer, whereas *field* is inclusive, hence closer to a quantum understanding of the relationship. Converting the notion of *scape* into that of *field* provides an opportunity to shift one's perception from the point of view to the point-of-being, or to a kind of existential participation leading to the quantum ecology. We could say that the characterizing feature of fields, as regions in which each point is affected by a force, is to give shape to a dynamic, either physical or virtual, ever in-the-making dimension. From here, we argue that today's increasing data saturation is prefiguring the emergence of a global datafield. We regard the datafield as the data-based and data-rich realm that can be associated pertinently to Appadurai's techno- and mediascapes, although the other three scapes, interpreted as fields, also inform the datafield through the very fact of being in constant development and integration.

We can look at the datafield as a field born out of the quantification of material and virtual entities. In the latent datafield, data are in an unexcited state, that is, they are virtually there, lying in waiting, available, but not yet summoned. They are there, in the first place, because people, institutions, companies, and governments produce them through daily digital-tech-driven activities, and yet, they have neither a priori specific relevance nor pertinence; they are potentially all equal. In other words, taken as it is, the datafield can be said to be in a state of isomorphic quietness in that data—whose production and circulation is by now paradigmatic—do not embed any specific use or meaning (which, however, does not mean to say that data are "natural"[2]). It is such isomorphism that makes the datafield extremely entangled in all its (f)actors. It is only when the datafield is summoned, only when algorithms are called upon to query *certain* sets of data in a *certain* way, that a *certain* pertinence emerges.

Once awakened, aggregated with other data, and brought to bear on the answer, data sets become relevant; by now, they are "excited." Data get hypercorrelated with other summoned data, *while* also loosening their entanglement with all that surrounds, which is made less pertinent. Put differently, the datafield—the quantified evidence—is in a state of global entanglement, which is then repeatedly reworked, loosened, and tightened, depending on how data are aggregated, made pertinent by/through algorithms, and put to work. Sure, one might claim that much data are *already* produced with a purpose in mind, but this does not prevent making them available, in principle, for whatever other purpose. It is only a matter of knowing how to tap into the datafield (technically and legally).

Hence, the availability of big data combined with the increased sophistication of algorithms and much accelerated computing power have set the essential conditions for datacracy to emerge. Data analytics, in that sense, is a global technocultural expression at the intersection of data (numbers), algorithms (math rules), and visual design (interface).

*   *   *

On February 16, 2012, *The New York Times Magazine*[3] featured the story of Andrew Pole, a statistician who, at that time, worked for Target, a US general merchandise retailer. His task within the company was to make use of Target's available data about customers' purchasing habits—gathered by giving each a unique guest ID—in order to come up with a model that could predict their future purchases so as to better tailor the retailer's marketing strategy. The targets of this mapping were especially young couples and women who would soon become parents and mothers, that is, customers with a potential long-lasting fidelity. In order to do so, it was necessary for Target to intercept soon-to-be parents at an early stage and ahead of competitors. This is why the retailer turned to Andrew Pole: "We knew that if we could identify them [soon-to-be parents] in their second trimester [of a woman's pregnancy], there's a good chance we could capture them for years," Pole said to the *NYT Magazine*. And this is what Pole actually did: by crossing purchasing data with various sets of demographic data concerning Target's customers, he was able to define a model that predicted couples' upcoming parenthoods. More than that, he identified twenty-five products that, when bought together (all or some of them), might accurately provide the estimate of a potential pregnancy still unknown to the future parents, thus opening the way to an advanced tailored marketing strategy. In fact, Pole's model was so accurate that its predictions also had unexpected consequences such as in the case of a family living on the outskirts of Minneapolis. One day, the father entered Target's store to issue a complaint with its managers because his adolescent daughter was receiving coupons and advertisements for pregnancy-related products. Target's managers apologized for the possible mistake. However, after a few weeks, it turned out that the young daughter was indeed pregnant, to the surprise of the whole family.

Target's example is a case in point for explaining the datafield. First, purchases are gathered in a guest ID and turned into sets of data (that are per se unexcited at this point); it is then Target's ability to connect—via algorithms—that data with other similar purchasing data and with precise sociodemographic details (other unexcited data) that allows the retailer to build its own predictive model, by capitalizing on the existing datafield. The predictive model is a sort of activation of the datafield, which enters a state of excitement and turns Target into a *data regime* (see further below).

This example is also significant in the context of today's digital transformation. First of all, it shows the extent to which companies are able to collect and triangulate *all sorts* of data, so that people's behaviors can eventually be monitored and predicted *even* without initially accessing the Web or digital devices. Second, it shows that such predictability has reached a high level of accuracy: indeed, it is so accurate that it anticipates people's life events and future decisions, even before such events and decisions enter their awareness. In other words, not only the predictability built on data analytics performs actions beyond humans' understanding from an operational point of view, but predictive analytics per se have become so efficient that they can easily bypass the subjects' conscious decisions. Data analytics operate at such a deep level that, based on whatever came before, they are now able to predict what will occur. More broadly, the Target example is symptomatic of the fact that by now, society lives in a data-saturated environment in which everything is susceptible to be turned into *valuable* data, regardless of people's awareness or active participation.

The global consolidation of the internet was the precondition to make the datafield technically global and isomorphic, that is, based on the same infrastructure so that data could be circulated. At the same time, it is on the Web that the datafield has found the fertile terrain upon which to blossom. On the Web, the datafield has found a golden mine, so to speak, in that the Web has created another layer of /reality/, a virtual doubling based *ex principio* on the digital logic of 0s and 1s. Such a further layer informs and is informed by the datafield in a mutual codependency that is literally creating a condition of virtualized purpose. This tendency of "redoubling" virtually people and things' experience, practices, and uses resembles the "cognitive aura" presumed even minimally of all living beings, including plants and bacteria. The ancient Greeks had a name for this phenomenon, which they called *entelechies*, thereby ascribing an inner purpose in animated organisms. Of course, there is no inner purpose in algorithms apart from the obligatory sequence that is imposed from the outside by the operating logic. This new virtualized environment, well past the Web stage, uses the datafield to summon an alternate space in such innovations as the metaverse, digital twinning, generative pretrained transformer (GPT), and now almost monthly emerging applications.

What the data-driven world has favored is the thickening of the datafield mantle (with increased possibilities of predictive analytics through big data) and its coverage of (almost) every aspect of life. At this point, the generation of data has spilled over into physical reality, for instance, through portable devices, so that the offline world has been as much engineered and datafied as the online world. Most important, this spillover produces a radical, qualitative change in the perception of what people presume is real and their relationship with it, as seen in chapter 2.

At the same time, however, it is important not to be too technologically deterministic. Despite such pervasive coverage, the datafield is neither an infallible nor a fully homogeneous resource. It is more of a quantified modularization of /reality/, which is uneven, thicker in some areas and thinner in others, and subject to resistances, as well as technical hindrances. An example comes from Google Flu Trends (GFT), a service launched by Google in 2008 (and active until April 2015) that aimed at mapping and predicting the fluxes of seasonal flu in twenty-five countries, based on the data coming from users' searches on the engine. In an article published in *Nature* (Ginsberg et al. 2009), the team behind GFT claimed that their service, together with the data coming from the US Centers for Disease Control and Prevention (CDC), could predict flu peaks two weeks earlier than the CDC's data alone. And yet, after a couple of years of calibrations of the service, GFT failed to predict the peak of the 2013 flu season by 140 percent. As explained by research conjointly conducted at Northeastern University, the University of Houston, and Harvard University (Lazer et al. 2014), the reasons for this failure are to be found in the method used by GFT to analyze data and, more specifically, in the algorithm at the base of the service that tended to overestimate correlations among data queries when, in fact, users often resort to Google's search engine not knowing precisely what to look for (a discussion that brings us back to the uneven relationship between data and information discussed in chapter 2). For the present discussion, GFT is a very apt example for showing that (1) the datafield is indeed widespread, but it cannot be turned easily into a omni-powerful system (in fact, it contains its own potential resistance), and (2) data cannot fully substitute or entirely map human behaviors, especially when these deviate from a rationalistic approach, when they are heavily biased (e.g., nonrepresentative), or when they confront uncertainty (not knowing what to look for) or psychological behaviors that contrast with the normative scenario in which algorithms perform (and which in turn contribute to create). In fact, Paul Edwards (2010, 84) talks of "data friction" to account for the technomaterial constraints that hinder the free circulation of data: "whenever data travel—whether from one place on Earth to another, from one machine (or computer) to another, or from one medium (e.g., punch cards) to another (e.g., magnetic tape)—data friction impedes their movement." Jo Bates (2018) goes beyond that by addressing three layers of data frictions: infrastructure and management, sociocultural factors, and regulatory frameworks. "Frictions," Bates (2018, 412) writes, "are constituted within complex and contested sociomaterial spaces in which various forces struggle to shape how data do and do not move between different actors." This is, after all, a very fitting characterization of the tensions running across the datafield and among its various actors (i.e., *data regimes*).

Data Regimes: Actualizing the Datafield

Within the datafield, different data regimes emerge. These data regimes can be considered "bundlings" able to inform, grab, and aggregate the datafield's basic resource—data—for their own purposes, be they institutional, commercial, political, or personal. Concretely, examples of data regimes are, among many others, Amazon and Alibaba; the US National Security Agency (NSA) and any other governmental body; digital collections and databases; social media platforms; and also (to an extent) users. The list could go on endlessly. More generally, data regimes are any node of data aggregation that inevitably reshapes the datafield with its own presence and interactions with other data regimes. We call them "bundlings" because these nodes are not "things," but techno-sociocultural aggregations repeatedly subjected to change. From the perspective of quantum physics, these data regimes can be thought as particle-like in that they are instantiations of excitement of the underlying datafield as a realm of possibility. Their "energy" comes from the algorithm—the true holy grail—that constitutes and defines their own possibility and mode of existence within the datafield. The more the algorithm capitalizes on existing as well as new data being produced, collected, stored, and (re)used, the more the data regime is excited. Clearly, data regimes behave and perform differently depending on the efficiency of their algorithms and the appeal of their services; it might well be the case that new data regimes emerge and consolidate, eclipsing older ones, simply because the algorithmic model of the emerging data regime resonates more with existing data and users' demands. A clear example is the rise and fall of Yahoo in favor of Google, the slow marginalization of eBay due to the preponderant role of Amazon, or the decline of a social platform like Flickr completely outplayed by Meta.

Most important, these data regimes can be more or less entangled with each other depending on the data they resort to and collect. This means that one can outline the entanglement of these data regimes—how much they are likely to simultaneously affect each other—by knowing their data interests (Hasselbalch 2020), the services they offer, the algorithmic models they use, and the users they aim to intercept. It must be noted that while the datafield is a nonlocal concept, data regimes are local: they give shape to technocultural (data)fields. Indeed, it is pretty obvious that data regimes, although capitalizing on the datafield, do not exist and act in a void, but they are rooted and/or have a reflection in the physical reality, within precise socio-political-legal contexts that determine their actions. The examples discussed below unveil the curtains over the entanglements that govern some of today's data regimes, be they public institutional actors or private corporate ones, either in the West or in the East.

*   *   *

There are jurisdictions current today, such as in Singapore or South Korea, where data analytics are providing sufficient and comprehensive information taken from the people themselves via the analysis of social media, smartphone contents, face-recognition cameras, and other data sources to justify policy and ruling decisions made for individuals in valuating, orienting, and positioning them for or against various benefits in education, housing, and health services. Security and privacy issues loom large in such practices and will be even more often invoked as the geopolitical as well as local safety conditions become more threatening.

In present-day China, guaranteeing security legitimates measures that clearly infringe on privacy such as ubiquitous face-recognition technologies to identify potential harm-doers or equipping Robocop-like police officers with direct access to criminal and other indicting records. China has taken a large step beyond such understandable policies (if not fully acceptable for other nations) by implementing the practice of giving "social credits" to individuals, based on the cumulative and permanently upgraded valuation of their behaviors and accomplishments or lack thereof in their daily life and over the long term. Keeping tabs on everybody is the accepted norm and has been so for millenaries. Although the subject of much exaggeration and misinterpretation, the fact is that social credits, whether restricted to reigning in the power of private enterprises or extended across the board to all Chinese citizens, is a coherent application of the digital transformation to a culture steeped into four millenaries of community rather than individualist preoccupations. It has been suggested that in a nation comprising almost a billion and half citizens, there would not be enough police to control the behavior of everybody and hence the move to guarantee the development of what could become a kind of "self-police state." And the increasing sophistication of data analytics that are well on the way to penetrate individual people's thoughts and motivations could lead to Orwell's dismal vision of "thought-police" but not necessarily imposed or implemented by a special government force but rather seeping through collective acceptance. MIT faculty member Yasheng Huang (2018) explains what makes Chinese people not only tolerate but also welcome the principle of social credits:

> One reason Chinese attitudes are different is that as recently as the 1980s, the word "privacy" had negative connotations in China. Chinese norms are anchored in 2000 years of a Confucian culture that values the intensity of interpersonal relationships. One way to solidify those relationships is through transparency and full disclosure. To be private was to be antisocial.

China's social credit systems create both a top-down (state–citizens) and peer-to-peer (citizens–citizens) bond. By doubling human behavior as automated tracked statistics, China's social credit systems fuel a moral competition among its citizens through a form of public surveillance: no longer sheltered behind screens (the third space), people are held responsible for any action and decision they take not only toward the state but also directly in the face of others. It is for these reasons that, in a society in which the fear of "losing face" is still relevant, the system is indeed popular and widely accepted. It rests upon a consensus that gives the impression to each citizen to be exposed to (and responsible in front of) the whole society. Via social credit systems, the behavior of the single person is by default reflected in the collective, producing a counterreaction of self-repression/self-control, no matter if for fear of peer censorship or of automated punishment. It is, indeed, a form of data-driven Confucianism.

This development seems to amount to a radical shift from the kind of self-censorship practiced by guilt-ridden Western societies to censorship by other people. Yet, it is not very different from what has been the norm in Western countries, since banking loans have been approved or denied for decades now on the basis of people's behaviors, assets, and careers. Bankers, lending institutions, businesses, human resources departments, insurance companies, and other administrations have openly attributed "credit ratings" to private customers, potential hires, and charges. Not to talk of the surreptitious, covert collection of people's data through initiatives such as that of the US NSA. Shoshana Zuboff (2019) gave it a name, "surveillance capitalism." Both surveillance capitalism and social credits show the way to the future, each according to their respective cultural ground, not primarily because of political maneuvering but essentially because the social body, rulers, and ruled alike are penetrated and transformed by the digital transformation.

*   *   *

An example of a major geopolitical entanglement between data regimes is the one that occurred in 2019 involving the US administration and Huawei company. On May 15, 2019, by issuing a security order, the Trump administration officially banned Huawei from selling its products in the US. Beneath this zenith point (or nadir) hides a long tail of events.

On December 6, 2018, at the request of US authorities, Huawei CFO Meng Wanzhou—daughter of Huawei CEO Ren Zhengfei—was arrested in Canada over allegations of having breached sanctions to Iran. This episode signals escalating tensions between the two parties, with the Chinese company being accused by the US of trade secret theft and fraud (January 19). At the center of the crisis is, at once, a commercial and political wrestling: the rollout of 5G networks around the globe led by Huawei, on the one hand, and the security (and commercial) implications that such endeavor could have, on the other. According to reports from US intelligence, the rollout of the 5G network based on Huawei technology would constitute a security threat, as it might lead to uncontrolled espionage. At the beginning of February 2019, the US government warned its European allies to prevent the deployment of Huawei's technology. Such a call received mixed responses: while, for instance, the Italian government temporarily stopped the rollout of 5G networks (end of February), in the UK—the first country in Europe planning to implement the Huawei-based 5G network—the risk of using Huawei's technology was deemed "manageable" (February 17); at the same time, in Germany, the US warning was perceived and publicly addressed by governmental figures as an undue interference (March 19). Lastly, on April 30, Vodafone reportedly found hidden backdoors in Huawei devices, which strengthened the US decision to enforce a ban on Huawei's technology two weeks later.

In the meantime, a number of US companies—for example, FedEx and Google—found themselves playing an in-between role as both Huawei's commercial partners *and* US actors bound to "national duties." Hence, while, for instance, Google cut off Huawei phones from future Android updates straight after the ban enforcement (May 19) and Facebook similarly prevented

Huawei from preinstalling its app on Huawei's mobile phones (June 7), by the end of June, it emerged that more and more US companies were bypassing the ban, with FedEx suing the US Department of Commerce, Microsoft still selling Huawei's laptops, and Google even warning the Trump administration that the Huawei ban would itself constitute a national security risk.

In some respects, however, the ban did affect Huawei's activities and strategy. The rollout of 5G networks was delayed, with countries in Europe calling out for a 5G security proposal (May 5); the company's revenues had to face corresponding losses, according to its own CEO (June 18); and, above all, Huawei was urged to accelerate the development of its own operating system. After denying and postponing this news, the Chinese company released its Android replacement—"Harmony," a buzzword with a political-historical resonance in China—on August 9, 2019.

More broadly, such a crisis is symptomatic of two different visions of the infrastructure that govern digital transformation and inform the datafield: multistakeholderism and multilateralism (Nonnecke 2016), with the former defended by those countries—mainly in the West—which, under the flag of a "free internet," advocate the need to maintain the digital transformation as deregulated as possible, so that actors and services can compete globally, while the latter, chiefly promoted by the BRICS and LMICs,[4] advocates the possibility of shaping bilateral agreements among states, as well as between states and ICT companies, thus fundamentally conceiving digital transformation as a sovereign matter. The significant thing is that, beyond discursive and pragmatic geopolitical oppositions, the data regimes in play are much more aligned than they (pretend to) appear. While US corporations dominate internet services and software (operating systems, search engines, social networks, browsers), the "ownership" of the internet infrastructure (submarine cables, content delivery networks, autonomous system numbers, internet exchange points) sees an imbrication of actors, with European countries and the BRICS (especially China) very much involved. To further problematize the scenario, the layout of internet infrastructures around the globe is led by hybrid joint ventures in which US, Chinese, and European actors are all playing a role. For instance, in 2013, Facebook acquired a $450 million share in the APG project—a system networking China, Japan, and several Southeast Asian countries—counting eleven other partners, among which, China Mobile, China Telecom, China Unicom, and Korea Telecom. One year later, instead, Google invested $300 million together with China

Mobile, China Telecom, SingTel, KDDI, and Global Transit to build the transpacific FASTER cable system between the United States and several cities in Japan, China, and Korea.

In fact, supposed competitors defending the two different visions to the internet, behind the scenes often find an agreement when it comes to the very basic principle of surveillance through data. As Iginio Gagliardone (2019, 149) notes, days before a conference in Dubai in 2012, which crystallized the two opposing camps for the sake of public opinion, "representatives at the WTSA were swift in passing a new standard on the 'Requirements for Deep Packet Inspection in Next Generation Networks,' or 'Y.2770.' Discussions happened behind closed doors and no drafts were circulated before a final decision was made, attracting criticism on the 'lack of transparency of the ITU-T in contrast to other leading global standards organizations.'" In other words, an agreement on the standardization of deep packet inspection (DPI) did find consensus among different actors, beyond contingent differences.

These cases are emblematic of the fact that the politics that sustains and develops the infrastructures at the base of the datafield is much more complex than the press releases reveal—and yet it is far from being globalist in a fair sense of the term. This is why, despite all ideological opposition, the deep trend is toward a federated multipolar scenario (Winseck 2017, 229) where (f)actors and discourses are very much entangled on contingent bases, reinforcing old power asymmetries as well as producing new ones from which no one can, in principle, be spared (Calzati 2020).

*　　*　　*

China's social credits and surveillance practices in the West are cases of political data regimes, while the crisis between the US and Huawei provides the scenario of an entanglement between (at least) two data regimes, one (Western) governmental and the other (Chinese) corporate. The third example discussed here has to do with a single Western corporate data regime, showing how it behaves differently depending on the context and the users, thus making different data sets and users/observers relevant to its performance. It is the case with Google Maps and its treatment of contested geopolitical borders around the globe.

On February 16, 2020, an article in *The Washington Post*[5] brought attention to Google Maps' practice of redrawing borders of disputed areas in the world depending on who is actually *looking* at them, that is, making use of

Google Maps from which location. So, for instance, Kashmir's borders, contested for decades between Pakistan and India, appear as fully integrated into India if the user is in India, while they look much more fragmented in case the user of Google Maps is in Pakistan. Far from being a mere technical matter, this and other similar examples are symptomatic of two orders of problems. The first one is the political value of any cartographic project: to map means not only to present a territory but also to somehow create it, in fact to build a vision of that territory with the more or less explicit goal to claim knowledge and take possession over it. Second, any cartographic endeavor can never be fully accomplished once and for all, or it would simply lose its relevance; it is rather an ongoing, temporary negotiation on how "things stand" at any given moment in history, and most important, such negotiation can be disregarded by one or more parties at any time.

It is then naively surprising, if not worryingly shocking, that Ethna Russell, the director of product management for Google Maps, declared (as reported by the *Washington Post*) that "our goal is always to provide the most comprehensive and accurate map possible based on ground truth" and that "we remain neutral on issues of disputed regions and borders." To mention "ground truth" and "neutrality" in matters of data mapping and borders not only is ingenuous per se but also, more problematically, hides the politics behind datafication. While "ground truth" is, as seen, a question of perspective and entanglement between viewer and viewed, "neutrality" sounds as an attempt not to take responsibility and depoliticize data. However, it is no longer solely the case that data (and technology) are neither neutral nor self-evident, but rather that they are able to impose a precisely engineered vision through their widespread application. The declared goal of Google is to "organize the world's information." Regardless of the supposed objectivity that this statement pretends to convey, this is an utmost political project, as the discord on borders' representation shows. *The Washington Post* points out that "while maps are meant to bring order to the world, the Silicon Valley firm's decision-making on maps is often shrouded in secrecy, even to some of those who work to shape its digital atlases every day," an admission that makes such a project even more problematic because it betrays the fact that, as data regimes tend to make the world more and more transparent and traceable, they tend to become more and more opaque and impermeable to public assessment.

Time and Space Smashed to Bits by Digitalization

After such examples of instantiations of the datafield and data regimes, let's dig deeper. What are the phenomenological characteristics of the datafield? We discuss two major dimensions: time and space, after which, we will delve into the individual–collective dimension. Needless to say—let alone from a physics point of view—time and space dimensions are very much interdependent.

Kairological Time and the Algorithmic Reversal of Time

Concerning the time dimension, the datafield is characterized by an always present condition that can be best understood by resorting to the Greek notion of *kairos*. Ancient Greeks had two main concepts to characterize time: *chronos* and *kairos* (although two more minor concepts could be identified: *aion*, which is the eternal time, and *eniautos*, which indicates a year time). *Chronos* is the concept that has become dominant today (and which was also the most prominent in ancient Greece), and it accounts for the sequentially, unaffected passing of time. It is a quantitative denotation that foregrounds the possibility of measuring time objectively. On the other hand, *kairos* identifies an intrinsically qualitative time, "the right moment" of/for a given action, or also "the time of the now," intended as the time that impresses the whole ethos to a given situation. For the Greeks, *kairos* could be discerned individually as well as collectively: it was the speaker and the audience who defined the relevance of a given moment as kairological. This makes *kairos* a subjective (but not solipsistic) and heterogeneous time, one that is perspectival and discontinuous by opposition to the continuous, objective chronological time. It is especially this idea of a heterogeneous time related to action, as well as of a time that is dependent upon an observer, that best fits the kairological time of the datafield.

As seen in chapter 2, data and the digital transformation inform and are informed by the logic of performativity enacted algorithmically. Data as a parceling of information is useful and datafication as a process is effective to the extent to which they can perform analytical and decisional processes as efficiently as possible, by maximizing the results given the available informational resources. In this respect, the time of data performativity cannot be measured in absolute terms, as it depends upon the context and on the outcome of the specific performance for which data have been summoned.

The right moment of the datafield is that which is functional and creates efficient performance.

This also means that the outcome of the performance must be assessed by the actors that have mobilized the sets of data in the first place. Observation is always crucial; data do not speak for themselves. Hence, the datafield is essentially kairological insofar as it depends upon the collected data that can be used at any given moment, from any given entry point, and for any prescribed goal. It is the performance that excites and brings the datafield alive. As the Target and Google examples show, data are made relevant by the performance itself, but this does not imply necessarily a positive outcome; in fact, each data performance constitutes a kairological moment of its own, which can resonate or contrast with those informing other data performances. This is why, in practical terms, one witnesses the blossoming of a variety of voices, such as in the case of the pandemic, each of which claims the validity of its own data analyses, from politics, to medicine, down to social sciences. In the age of the datafield and data regimes, there is no longer (or not yet) an overarching time–space frame able to legitimate a commonly accepted data hermeneutics. The misunderstanding originates in the conflation of science as a method and data as an outcome. On this point, it is worth referring to Linnet Taylor's work analyzing tracking systems' failure during the Covid-19 pandemic. Notably, by calling for a shift from monitoring to caring, Taylor (2020, 4) writes that "building the capacity to measure and track have often been emphasized over the ability to understand exactly what has to be measured and tracked." Such a statement calls directly into question *what* data can really represent, *how* they do that, and *why*.

In this respect, the kairological time dimension of the datafield is precarious in the same way in which Heidegger (2004) conceives of *kairos*, notably as a fundamentally insecure time, a time that does not unfold but simply happens, like an epiphany. Heidegger roots his discussion on *chronos* and *kairos* in theology and in particular in the hermeneutical reading of the letter of Paul to Thessalonians. According to Heidegger, a paradigmatic example of *kairos* is found in Paul's description of the *parousia*, the second coming of the Messiah. Indeed, Paul conceives of the *parousia* as a moment that people cannot know in a chronological perspective; rather, the second coming will happen unexpectedly, at a time and in a form they cannot foresee. There is no sequential time at all in *kairos*, only emergence.

From our perspective, this means that the time of a given data performance reverberates into the whole datafield in ways that are not chronological but kairological: every data performance is and configures a whole new scenario at every (new) moment, at every (entry) point. People are confronted with an endless precipitation and regeneration, an ongoing traversing and reconfiguration of the datafield. Now, what does this mean concretely? For one, that the "explainability" of data-driven processes might not be that simple to achieve and preserve. We are talking here of a cornerstone principle in many documents dealing with AI ethics and digital transformation, according to which humans should always be in a position to understand how data-driven decisions are taken by the AI and, if needed, to challenge/redress these decisions. However, what happens is that data performances define at every point in time their own aggregating-recombinant logic, and at the same time, they reshape the whole datafield. It is not simply a matter of density of networks or of flow of data but rather of a "leap" in the kind of synergies and outcomes made possible by the increasing entanglement of data regimes among them and with regard to the datafield. Such synergies create data performances, the outcomes of which are unique, that is, marked by a contingent time–momentum that is hard, if not impossible, to unpack.

The kairological time of the datafield, then, is a potentially ever-present time, a time that has no projection beyond its mere happening. Such temporality—which is inherently *uncertain*—also runs through the datafield insofar as it cannot be really known, from a theoretical point of view, how data will be used, for which purposes, and with which outcomes. The paradox is that while the datafield springs out of the quantification of every aspect of life, its own time (and space) cannot be measured, as this is intrinsically discontinuous, point-like. In this respect, scholar Byung-Chul Han (2016, 30) acutely argues that the time connoting the digital transformation "has no anchor, consisting of points that vacillate without a direction, creating both the tyranny of instantaneousness and a discontinuous present." According to Han, people increasingly perceive time as fragmented not because it is accelerated, but rather, time appears accelerated because things and objects have become disposable by being technologized. And the major consequence of that can be subsumed under two major trends: the commodification of memory and the erosion of imagination.

We have already encountered the work of Leroi-Gourhan. In *Gesture and Speech*, he (1993, 237) contends that "the operational synergy of tool and gesture presupposes the existence of a memory in which the behavior program is stored. With animals, this memory forms part of organic behavior as a whole, and the technical operation becomes, in the popular sense, 'instinctive.'" As soon as the tool becomes a technology and, detached from the limb, is able to perform a task on its own (such as, collecting, storing, and retrieving data), then memory too, as a program, is increasingly shared between the individual and the technology. Looking, for instance, at social networks, they have been very effective in turning daily narratives into disposable moments that can be recollected at any time by either the individual or technology itself (for example, through weekly reminders). In doing so, the technology produces a particular kind of memory, which might be called "stock memory." This is a concept that echoes the well-known notion of "stock images" as pieces of visual information that are "generic" and whose core features reside in "their alienability from a particular referential source, their autonomy from a specific intentionality of use and reception, and their archival origination" (Frosh 2013, 134). On social networks, memories, however personal, become interchangeable; people relinquish to technology the possibility of authoritatively assembling them. Far from defending here an essentialist conception of memory—one never remembers in isolation—at stake is the idea that once people forfeit the privilege of remembering (as a process) and delegate it to technology (as a product), this opens the way to an automated homogenization (and impoverishment) of what and how people (are supposed to) remember. The point is that, as Clive Thompson (2013, 33) remarks, "Even if we are moving towards a world where less is forgotten, that isn't the same as more being remembered." Technology does not know oblivion, nor can it conceive of (un)conscious removal: its horizon is that of a depthless, ongoing present, where everything that is manifested is also relevant. From a Freudian perspective, one could say that technology does not remember (*Erinnern*) events but only repeats (*Wiederholen*) them because it does not really work through (*Durcharbeitung*) memories, but only stores and relaunches them following a quantitative (meaningless) logic.

Concerning imagination, according to Harari (2016), humans are the sole species able to cooperate in large numbers and flexibly, that is, coordinate complex patterns of behavior, as well as adapt these behaviors and create new ones depending on the circumstances. Hence, people are able to make complex plans for addressing, in advance, future issues, which means to be able to not only get collectively organized but also foresee how certain scenarios will evolve and how people's organization might be able to tackle them.[6] Stuart Kauffman and Andrea Roli (2022, 4) write in this regard that "discovering a useful but complex sequence of 'actions' is a blind search in an indefinite space of possibilities. . . . Creativity is not deductive, it is insight. . . . We cannot find new features of the world by deduction, induction or abduction. Insight is required. Insight is not deductive." While the authors contend that this is what makes humans unique, the latter's imaginative supremacy might soon be contested not by other animals but by the emergence of increasingly powerful and generative forms of AI of which humans already hardly understand the functioning and performances. Humanlike intelligence or imagination should not be the only yardstick to judge the data-based performance of machines. For instance, in 2017, *AlphaGo* computer defeated the game world champion, while a more recently developed algorithm was able to play the complex *Starcraft* game, pointing to the fact that machines' affordances cannot be evaluated through anthropological lenses solely.

In an article published in *The New York Times*,[7] McLuhan introduced a stunning insight about the resonance between Freud's dream theory and that of quantum mechanics:

> The twentieth century opened with Max Planck's theory of quantum mechanics in 1900, stating the discontinuity of the material universe. In the same year Sigmund Freud published his *Interpretation of Dreams* stating the discontinuities of our conscious and unconscious lives. So far as I am aware, economists have not yet matched physics and psychology with any statement of the discontinuity of the economic bond. All existing theories of inflation are hardware theories, nuts and bolts theories, theories of connected and continual rational processes of supply and demand. The equilibrium theories of supply and demand concern the quantities of "hardware" as it were, whereas the disequilibrium realities occur at the speed of "software." "Software" is the world of electric information and also computer programming. All of these constitute a new service environment of electronics pulsation which makes possible the dealing in "futures" and the anticipation of the gaps and intervals in supply and demand.

Through hyperconnectivity, people are indeed already living ahead of themselves in uncertainty and indeterminacy. This throws society in a perpetual forward movement, gambling on the future. Thus, the digital transformation has changed the vector of time, overtaking the unfolding present to bring people into an immediate (unpresent) future.

Today, because they resort to increasingly responsive and adaptive tech- nologies, people's imagination is further decommissioned. Every time users rely on Amazon or Netflix recommendations, they suspend their imagina- tion—as a form of self-reflection over what is expected and desired—letting technology take over and do the job. Suspension and delegation obfuscate the possibility for people to (fore)see themselves in the world—both individually and collectively—as a form of *dasein* that requires and stimulates insights and open-ended awareness and responsibility. In other words, the digital trans- formation undermines humans' capacity to act collectively in view of future scenarios (of which the inanity before climate change is only one example).

Sure, one could argue that such forms of suspension and delegation of human imagination have always existed; just think of traditional marketing strategies. After all, this is how the human brain works. The aforementioned article about Target on *The New York Times Magazine* mentions that "one study from Duke University estimated that habits, rather than conscious decision-making, shape 45% of the choices we make every day." The point to make here is that, nowadays, through platforms in which decisions are always one click away, the margin of decision has shrunk tremendously. People have been turned into hyperhabitual beings (i.e., quasi-machines) and there is increasing evidence (cf. Libet 1999) that "instinctive decisions" light up in the brain even before becoming conscious.

According to Stiegler (in Lemmens, 2011), every technology leads to a *"proletarianization"* of the subject, insofar as every time people interact with it, their abilities and faculties are eroded, by forfeiting a portion of their know-how (what he calls *savoir faire*) to technology itself. This highlights the extent to which using technology means, at all times, also to be used by technology; tech smartening and human stupidification go hand in hand. Smart technologies leave users wondering *why* they did what they did; these technologies conceal their own complicatedness and, in turn, suppress the possibility of alternative behaviors, desires, and expectations from the user side, reducing and formatting the user's complexity for the benefit of compu- tation. It is in this respect that Calzati (2018, 221) speaks of hackers as poten- tial barbarians within the digital ecology: "The hacker not only understands the backstage of digital technologies, but has made that dark area its front stage. The hacker is a barbarian who knows the language of the 'doxa'—i.e., the code with which digital technologies' algorithms are written—and uses this language heterodoxically, that is, against technology itself."

Space Fragmented

As seen in chapter 2, people now occupy three spaces. There are, how-ever, things to be aware of when considering the relationships between the three spaces. Although people occupy those spaces simultaneously in different proportions, each one has effects and impacts on the other two. While one can assume the relation between the mental and the physical spaces, the third space dislocates and separates the other two. Different from mental and physical experiences of space that are perceived as con-tinuous, people's relationship to virtual space is constantly interrupted and delocalized. There are cases, however, where the virtual space affects the physical one directly in a supporting manner. A good example of this rela-tionship is the rapid rise of what is called the industrial metaverse where virtual shop floor activities aid or supplement physical labor and materi-als.[8] Another example, proposed by de Kerckhove, would be the concept of the "metacity," now rebaptized "cityverse" by the European Commission.[9] In de Kerckhove's view, the metacity would combine the administration, entertainment, and industrial uses by re-creating the physical environment in virtual format and, for example, allow residents to entertain guests glob-ally in their own decor or help the urban managers to plan /real/ events or new cityscapes in exact virtual simulations.

Today, the datafield is characterized by a shrinking of distances down to a sort of collage. The physicist Carlo Rovelli (2020) thinks of space as literally particularized. This idea can be best understood as the radicalization of what Luciano Floridi (2017, 123) calls the "cut-and-paste logic" at the basis of the digital transformation: "The digital cuts and pastes reality in the sense that it couples, decouples, recouples features of the world." Such logic fragments space to the extent this latter becomes point-like, such as in a pointillist painting; at the same time, as in a painting, the sense of three-dimensionality is very much there. In fact, this is more an ongoing (and endless) process of fragmentation, leading to ever new forms of recombination of colors and shapes, information and data, rather than a definitive pulverization of space. At stake is no longer a—by now purely analytical—distinction between online and offline, which are two fully codependent realms, but the binary formatting—necessary for subsequent computation—of /reality/.

This logic brings consequences. For instance, the shrinking of space implies a virtual global deterritorialization *and* consequent ever-localization. The user is increasingly enmeshed in the datafield and simultaneously entangled with various data regimes. In fact, most of these processes do not require, by now, the user's overt consent; it is simply sufficient for the user *to be there* in order to be framed by data-driven technologies as a data subject and become part of the datafield. Most important, this *being there* of the user is actually to be *anywhere*, meaning that there is a scission between the subject's virtual and physical point-of-being. This is why people are often surprised by the concrete consequences that their words and actions online have on their physical lives. It is a tech-based dispossession.

Data configure a discontinuous point-like space, made of aggregates, which can be constantly redefined and enacted as needed. In turn, this also implies that the way in which information reaches the subject predisposes a fully integrated form of assimilation (subject-as-target), which hinders active acquisition, due to the ever-expanding and ever-denser networks in which the subject is caught.

*   *   *

The fragmentation of space is manifested in the increasing proliferation of governmental e-services—from health care to fiscal services down to IDs and voting—that literally detach the subjects from their physical place. The country of Estonia is certainly a case in point in this respect. Since the early 2000s, Estonia has heavily invested in digital infrastructures and projects, which have become, rapidly, the backbone of the country's economic, social, and political systems. By now, Estonian citizens and residents can access 98 percent of governmental services online. An apt example for the present discussion is Estonia's e-residency program. The project was initially launched in 2015 in a beta version with the goal to provide citizens outside of Estonia with a digital ID and the opportunity to run location-independent businesses *as if* in Estonia, that is, by capitalizing on the country's advanced digital infrastructure and services from anywhere in the world, as well as on the EU's economic market as a framework.

The growth of the service was accelerated by the launch of an online e-residency application form: while, at an initial stage, e-residency applicants were required to either physically go to Estonia (Kotka et al. 2015) or visit an Estonian embassy abroad for completing the application process by proving their identity, fingerprints, and the payment of a 100 euros fee, subsequently the process has been fully digitalized, only requiring to choose the closest pickup point to collect the e-residency kit once the application has been processed (this remains, however, a major hindrance to the program).

Hence, e-residents are bestowed with a (location-independent) digital citizenship. However, in order to make this possible, they have to become "digital subject" in the first place, in that they have to be subsumed under a data-driven system in order for the service to be effective. More broadly, this is a good example of two simultaneous "decouplings," as Floridi (2017) calls them, made possible by today's digital transformation: that between location and presence, on the one hand, and that between law and territoriality, on the other. In fact, Estonia's e-residency program is an entangled phenomenon that emerges from the splitting of these binomials and leverages on their recombination. For becoming an *Estonian* e-resident (the adjective rather than the prepositional phrase "in Estonia" is crucial here), the monadic fusion of presence and location is not required, insofar as any subject can potentially apply to the program from anywhere in the world. Presence, then, is just "presence in the world," and any entry point to the program is, in principle, deemed valid. Similarly, the Estonian law comes to extend beyond its physical territoriality, and it does so by being applied to a digitalization of both the subject—indeed, a data subject—and physical placedness, which sublimes into a nonlocal space, in effect the datafield. It is significant to remark that, to the extent to which the e-residency is built upon the recombination of these binomials, the program also produces an entanglement of its own, notably that between the (data) subject and a set of actors—banks, public authorities, other e-residents—with which the data subject inevitably gets enmeshed.

Digital Twin and Smart Cities: The Entangled (Urban) Self

The digital transformation also has a profound impact on personal identity and one's existence into the collective. Two key configurations of the digital ecology originate at the intersection of the individual subject and the ever-responsive environment: the personal digital twin and the smart city, evolved into the city digital twin.

Personal Digital Twin

One of the most important effects of the digital transformation on one's own self is the emergence of a "personal digital twin" (PDT), an augmented doppelganger that is quite literally entangled with one's being and acting in the world. The term *digital twin* was first coined by Michael Grieves ([2002] 2014) to mean a virtual/digital representation of any physical object. Engineers used it first to designate the digital replication of costly motors and installations such as turbines and rotors, the idea being to facilitate their real-time monitoring and management. A machine's digital twin includes not only real-time reporting on its functioning but also the history of its maintenance, that of the occasional breakdown, as well as the source and coordinates of all parts' suppliers and records of delivery and efficiency. In many cases, the digital twin enables automatic repairs just as it regulates normal functions. More than a replica of the object's life cycle in the digital space, the digital twin expands to achieve a life of its own that is kept in sync with the physical twin. In fact, differently from traditional modeling, in order to have a digital twin, an existing physical object and its digital simulation must be fully integrated *in both directions*, so that "a change made to the physical object automatically leads to a change in the digital object and vice versa" (Fuller et al. 2020).

From the industrial sector, the term soon migrated to cover digital duplications of business operations, institutions, and labor forces. Eventually, it also applied to one's own self. Through data, the subject becomes a fully transparent formation; more than that, such formation can be replicated, distributed, and extended beyond the subject's physicality, by aggregating one's own dispersed data into what becomes a virtual sublimation of oneself. We have called "digital unconscious" all the data strewn about everyone in the world's databases. The individual subject might find in the digital twin a critical reversal of that dispersion. Suddenly, the unconscious coalesces into an identifiable formation, which becomes accessible to consciousness. This could have already been partly done by finding a way to integrate and use all the information contained in one's smartphone, but new GPT platforms, such as ingestAI, are simplifying the process by combining that data sourcing with whatever can be found online about the user. Thus, the digital entity becomes a parallel manifestation of the subject, an eruption at the conscious level of the underlying layers of information in a new unified instantiation that more than represents, actually engages with the physical person, since they mutually coevolve. The PDT is neither fully private nor exclusively controlled by the subject; it serves as a bridge between one's physical being and the datafield; better, the PDT is one's own data regime, emerging as a consequence of the subject's codependency—willingly or not—with the datafield.

Most advanced medical services are already heralding a new era of personalized medicine based on the twin's archives of the subject medical history combined with a lifelog, that is, the recording of everything the physical counterpart has done over the years, having access to varied sets of data, combined with increasingly refined analytics.

What are the consequences of the emergence of the PDT? Over and above predictable ones such as issues of privacy, hacking, identity theft, market co-optation, and so on, there is the profound anthropological mutation of one's sense of self. The self is emigrating: "for sure, we know that the self acts simultaneously as the separation and meeting point (between the subject and the outside world), and as such it allows us to establish relationships (between the I, the others, the world)" (Iaconesi and Persico 2017, 89). While humans may still believe that they have a self that is situated somewhere *inside* their body, the actual fact is already beyond that point. Indeed, the digital twin represents a cornerstone in the digital ecology taking over the language one. While literacy created the internal self, the PDT fully materializes its coalesced externalization. And this externalization will become increasingly refined and autonomous. One may take for granted to be a free agent (whatever that might still mean today), but *datacracy* already crystallizes its own epistemological horizon. The internal self will not disappear entirely but may become subservient to the external/augmented one; in fact, the latter will impact and reflect on the former as much as it is informed by it. Most important, that is not something people should leave fully in the hands of governments and corporations. Hence, to be in question are the very notions of transparency, accountability, and liability whose (language-based) normativity might no longer work, requiring a revision from an entangled data-physical perspective.

The digital twins will likely become everyone's interface to navigate and negotiate our relationships with and within the datafield. Hence, the one urgent question about it is: How autonomous will it actually be? Or also: What/where will the people's point-of-being be when dealing with their PDT? Should people grant to the physical body a sort of prominence, even if and when one's agency is fully aggregated in the datafield? Today, responsibility is (still) fundamentally based on an essentialist conception, that is, one that is identified within individual (physical or juridical) beings.

It is likely to witness a transition toward "distributed responsibility" that encompasses biological, digital, and legal actors. Yet, this distribution does not concretize in the form of a network, but an entanglement. In fact, data regimes already incorporate multifaceted techno-bio-social agencies. This requires new approaches, for instance, to define both a "right to die" for (the agency of) the human (as a biological being) and a similar "right to obsolescence" for technology itself (and here the social actors responsible for its production come into play with both rights and obligations). Or also, to counteract datacracy, it is necessary to legitimize the possibility to "remain analogue." Persistent connectedness should bring with itself checks and balances in the form of a legal framework that allows one to prevent the responsive datafield from arbitrarily tracking data. In other words, it is fundamental to grant the right to opt and act out of the datafield, without being denied access to fundamental services. In 2016, the United Nations formalized the idea that access to the internet should be considered a human right, thus calling for efforts to make it as widely open and inclusive as possible. This formalization came as a recognition that by now the internet and ICTs in general hold a huge potential for people's emancipation by allowing them to get information, have access to opportunities, foster a network of contacts, and so on. After all, it makes sense to imagine that, while schooling was made compulsory around the world as a way to educate generations to literacy, the open access to ICTs constitutes today the necessary but not sufficient (see second and third digital divides) precondition for any possibility of learning and making sense of, if not understanding, today's digital ecology. Any other position, in this regard, would be brutally discriminatory.

Nonetheless, we contend the need to complement the UN's formalization with a symmetric one that maintains the system in balance by making sure that it does not turn into a one-dimensional technocentric paradigm. It is vitally important for society as a whole to accompany the free access to digital connectivity as a human right, with an equal right to remain analogue (cf. "data justice"; Taylor 2017). While some may claim that this double-track scenario could lead to a waste of resources, we argue that a fully datacratic system, with no exit strategy, with no alternative frame by which it can read and interpret itself, is doomed to be totalitarian. The copresence of at least two equally viable and complementary paths is what makes an ecology healthy from a democratic perspective. Gilles Deleuze (1995,

129) correctly noted that "repressive forces don't stop people expressing themselves but rather force them to express themselves." To consider the access to connectivity as a right is essential, but to consider such access as the only way to obtain certain services is a forced imposition. After all, it is not because a literate culture imposed itself that educators stopped teaching how to deliver effective speeches. The right to get connected cannot be conflated with a (disguised) duty to do so: this means, simply put, that the (politics of) digital ecology needs to be secularized in the same way as language-based power and authority have been since early modernity.

From Smart Cities to City Digital Twins

> In that Empire, the Art of Cartography attained such Perfection that the map of a single Province occupied the entirety of a City, and the map of the Empire, the entirety of a Province. In time, those Unconscionable Maps no longer satisfied, and the Cartographers Guilds struck a Map of the Empire whose size was that of the Empire, and which coincided point for point with it. The following Generations, who were not so fond of the Study of Cartography as their Forebears had been, saw that that vast Map was Useless, and not without some Pitilessness was it, that they delivered it up to the Inclemencies of Sun and Winters. In the Deserts of the West, still today, there are Tattered Ruins of that Map, inhabited by Animals and Beggars; in all the Land there is no other Relic of the Disciplines of Geography.

This short story by Argentinian writer Jorge Luis Borges, titled "On the Exactitude in Science," is a case in point for unpacking the epistemological limitations of any spatial mapping, an endeavor that is also at the basis of today's "smart cities" as projects for "utopian 'clean and orderly' pervasive computing" (Viitanen and Kingston 2014). As Polish engineer and philosopher Alfred Korzybski put it very sharply, "the map is not the territory"; in fact, any model is useful to the extent it is able to offer a *certain* level of generalization of what it pretends to represent. The etymology of the word is clear in this regard, since *model* comes from the Latin *modulus*, which means "measure, standard." Hence, a model is the representation of a physical dimension (or phenomenon), which is based on a set of criteria for abstraction that have been agreed upon (i.e., socially and/or scientifically validated) and that are coherently organized in order for this representation to serve a purpose.

On the one hand, what Borges's short story does is to unveil the pretentious and ultimately failing attempt to map and take control over the whole world via a purely rationalist (measuring) approach. The outcome would simply be "unconscionable." Any model (even the model that science provides) is just a peculiar formalization of physical reality, that is, an (approximated) understanding of reality that is valuable under certain conditions of truth. It is significant, in this regard, that the title of Borges's short story is "On the Exactitude *in* Science," which can be understood as "On the (Impossible) Exactitude *of* Science." Science is always partial, incomplete, or, as Bohm ([1980]; 2002) put it, an "insight."

On the other hand, the story shows that even if people were to map all physical reality, this would *not* amount to better knowledge. The more is not necessarily the better. As seen in chapters 1 and 2, knowledge is a matter of not only granularity but also embodiment. The unmappable—all that resists a certain modeling—is what makes knowledge (of the mappable) worthy and useful. What is given (i.e., *datum*) is just a segmentation of what people are able to think and grasp from within a certain enframing *dispositif*.

*   *   *

The concept of the smart city has a longer record than that of the PDT. This notion, indeed, can be traced back to that of "wired cities" (Dutton et al. 1987) in the 1980s, followed by the idea of "digital cities" (Boschert and Rosen 2016) in the early 1990s, which was later coupled with that of "smart growth" (Bollier 1998) since the late 1990s. For one thing, this genealogy is symptomatic of the technological evolution of the past four decades—today, no one would think of wiring a city—as well as of the emergence of environmental sensibilities.

Coupling the concepts of "digital twin" and "smart city," today municipalities around the world are embracing the idea of a city digital twin (CDT). A CDT is a three-dimensional dynamic model that can help synthesize data from various sources (e.g., geospatial information systems, internet of things [IoT], archival data, social media) to create an integrated real-time knowledge of the city, as well as scenario simulations, both in the short term and long term. As such, CDTs bear high expectations from tech experts, city officials, and policymakers (Shahat et al. 2020) as a tech-driven solution to tackle the complex problems affecting cities (Bettencourt 2015). To achieve that, Simon Elia Bibri (2018, 238) writes that "regardless of their scales, new sensing and computing devices are projected to be equipped with quantum-based processing capacity, unlimited memory size, and high performance communication capabilities." Examples are already there: IBM, for instance, has developed all-round platforms called "Watson IoT" and "Qiskit," which allow, the former to manage large-scale data systems in real time, by also providing cloud and blockchain services, and the latter to run quantum simulations and cloud-based quantum processors.

When smart technologies and digital twins are implemented in the context of the city, intertwined issues of modeling design and governance emerge. Paralleling Borges's story, at stake is the fundamental acknowledgment—and consequent operationalization—that any process of (digital) mapping always provides a *certain* formalization of the phenomenon it stands for: while this is what makes the process heuristically useful in the first place, it inevitably entails a "translation" from the physical to the digital bringing epistemological and ethical pros and cons that demand ongoing finetuning. Far from constituting a mirror of the city, a CDT delivers *one* possible modeling of urban phenomena, behaviors, and spaces, which depends on tech affordances—*what* the used technologies *can* grasp and *how*—and non-technical aspects, notably the design and use (by whom and for which purposes) of such model (*why*). Yet, most of the time, in discourses surrounding tech innovation *in/of* the city—rather than *for* the city—technology takes it all: it is not rare, indeed, to hear possible socioeconomic or also environmental shortcomings being downplayed by heralding technology as an all-disrupting *and* all-fixing driver.

However, how, where, and which smart technologies are deployed in the urban environment can contribute to augment socioeconomic gaps, as well as create new ones: Jenni Viitanen and Richard Kingston (2014, 811) warn, in this regard, that "inequality and poverty do not often feature in smart city debates, but the technological fixes in smart cities will have distributional consequences under which there are winners and losers." Moreover, far from being green and sustainable, smart cities are likely to negatively impact the environment and environmental justice. Already in 2004, a study conducted by Lorentz Hilty and colleagues (2004) evaluated the impact of increasing pervasive computing both on the environment and on people's health, concluding that such tech solutions increase the risks for health and bring ecological setbacks, often in the form of unaccounted rebound effects.

Beside this, tech solutions adopted in cities are often co-opted by market forces and actors, which then define the normative idea(l) of what "smart" means (i.e., something that can be profitably benefited from). In fact, the underlying rationale of cities' smartening and twinning is economic performance, not human and social thriving. It is not surprising that the idea of "easiness of use" gets conflated with that of "living well-being," overlooking, for instance, what Richard Sennett (2018) calls the "stupefying effect," that is, the fact that the overreliance on technology leads citizens to suspend their critical sense. Fabian Dembski and colleagues (2020, 3) summarizes this point, stating that "the smart city as a product for the rationalization and technologization of the city becomes a neoliberal product." In fact, smart cities are one of the most profitable markets created by/through digital transformation. According to recent stats,[10] the overall market value for smart cities will surpass $2.3 billion by 2025. Most important, tech innovation is led by mostly private companies, leading to a subordination of the public sector to the role of recipient or client.

Third, it is not rare to find cities' smartening and twinning agendas coupled with the idea of "safety." A safe city shall be intended above all as a "monitored city," along the line connecting "care" and "control" (Lyon 2007): the extent to which safety is inflected more in terms of the latter rather than the former is an economic-political matter. As Deleuze (1992, 4) acutely points out by distinguishing between an old and a new kind of city governance, "The disciplinary man was a discontinuous producer of energy, but the man of control is undulatory, in orbit, in a continuous network." Today, one is a citizen to the extent one's behavior can be tracked, collected, modeled. Here the transition from *being* to *behaving* emerges in all its complexity, bringing with itself a denotation of the urban environment: "Any one of us," Benjamin Bratton writes (2016, 152), "is (or could be, or should be) less a political subject of this one city—London, Mumbai, Shanghai—but of the City, of the globally uneven mesh of amalgamated infrastructures and delaminated jurisdictions." The ICT-supported city is an agglomeration that transcends borders to impose a new abstract layer of global technologization in which the citizen is primarily thought of as a *citizer* (citizen + user). The *citizer* (not necessarily a human being) is both a wave and a particle: a continuously trackable (digital) wave and a (digital) particle whenever people actively (or passively) make use of the interfacing affordances made available to them by the city's datafield. In the near future, *nonsmart* citizenship might become a privileged locus of resistance against the increasingly pervasive datacracy: as Bratton (2016, 159) correctly notes, "We will find that in the future, the noncitizen may in some ways enjoy certain advantages over the citizen as infrastructures may not already be preprogrammed to govern that User directly as a formal subject but merely to transact services with her."

It is therefore necessary to reconceptualize smart cities and CDTs to counteract the predominance of technology (and private actors). This entails conceiving technologies *for* the city, that is, as being part of a sociotechnical dimension by which they are informed and which, in turn, they contribute to inform. As Timea Nochta and colleagues (2021) point out, "The usefulness of CDTs in decision-making depends on the success of reframing high-level policy goals into practical policy problems to which the model can suggest solution options. This reframing exercise must be informed by in-depth local knowledge and preferences and thus requires a participatory approach." This goes in parallel with an understanding of cities as agglomerations that cannot be approached as a machine (Mattern 2021), that is, as something to be broken into smaller parts, and then processed and recombined. Rather, cities are "hybrid complex systems" (Portugali 2011) composed of biotic and artificial elements, whose entanglement identifies a unique ecosystem. As such, cities cannot be studied by isolating either elements or their interactions but require to be studied in their entirety, insofar as they manifest emergent behaviors that cannot be fully predicted (Grieves and Vickers 2017). On this point, Luis Bettencourt (2015) specifies that it is self-organizing practices—that is, practices able to give agency to the needs of all actors involved—that represent the best response to tackle a city's complexity, by "placi[ing] emphasis on creativity and on effective social organizations, capable of coordinating their knowledge and action." Hence, a governance of smart cities and twinning models shall be able to move toward "more extensive public consultation, collaboration and coproduction" rooted in "a set of civil, social, political, symbolic and digital rights and entitlements" (Cardullo and Kitchin 2019). Technology can be useful as long as it coordinates urban coproduction, while it becomes a hegemonic tool as soon as it impedes the very coexistence of irreconcilable positions, behaviors, and visions.

*   *   *

Above all, it is necessary to advance the idea of CDTs as *urban digital twinning*, thus acknowledging that the implementation of smart technologies and the design of twin models are not reified solutions but in-the-making and always-partial processes that keep people at the center (Calzati and van Loenen 2023c). Such sociotechnical procedural understanding allows one to explore the entanglement of users and resources; people and data. Virtuous examples of urban digital twinning exist. Dembski and colleagues (2020) write,

> The Herrenberg digital twin differs from other simulation-based studies in the field of smart cities, in particular by linking and combining various urban data from models, analysis, and simulation and by the implementation of social data collected from citizens. Furthermore, the visualization in virtual reality ("virtual twin") not only enables broad citizen participation but also collaboration between stakeholders.

Herrenberg's digital twinning is led by the city's municipality. In this respect, the readiness of public authorities to take the lead in the twinning process is crucial to guarantee that the city becomes not only smarter but also more sustainable as far as the technology employed and the solutions deriving from it are concerned, thus subjugating private interests to collective ones. Moreover, a project designed in this way facilitates consensus building through forms of citizen participation, while putting social, economic, and environmental issues on the same level. As soon as citizens are involved and informed, they can actively take part in deliberative processes, fostering a new cultural affiliation that is meta-procedural with regard to the issues being discussed. Forms of local, iterative epistocratic decisional processes, whereby the right to deliberate is inscribed into a virtuous circle of knowledge acquisition and dissemination, can also be devised. Finally, to consider and enact urban digital twinning can also help to rescue technology from its doomed fate—as Zuboff (2013) points out—of being repeatedly turned into a tool of control and subjugation. The very notions of participation and transparent accountability get thickened and inscribed into a looping practice in which technology is both de-essentialized and de-commoditized toward dynamic processes of co-innovation and responsible research and innovation.

Symbiotic Autonomous Systems

The necessity—advocated in various policy documents, including the ethics guidelines on "trustworthy AI" of the European Commission (2019a)—to maintain human oversight over data-driven technologies and AI implies an understanding of humans and technologies as fully distinct entities. In this way, the monitoring of the functioning of the latter is conceived as a linear and transparent process performed by the former. However, things are not so simple. The symbiosis between (increasingly responsible) technologies—data regimes' performances in context—and the people being present and part of the datafield is profound and likely inextricable (this will be exacerbated by/through quantum computational systems because "running the same program over and over again will not necessarily lead to identical output" [Perrier 2021, 11]).

Warnings have been launched from various experts concerning the risks to embody AI into anthropo-looking androids, which might contribute not only to further blur the distinction at an apperception level but also to develop forms of empathy from humans to machines. Along these lines, a 2019 IEEE White Paper[11] talks of "symbiotic autonomous systems" to herald a new technohuman dimension:

> We are starting to see the emergence of a Digital Age in which the material to be manipulated is no longer (just) atoms but also bits. We are entering into this new age through a symbiotic relationship with our digital tools. These new tools have become complex entities that are probably better referred to as systems as they are starting to operate autonomously, due to a growing flexibility and awareness.

This means that AI decisions are, and will increasingly be, the result of entangled performances among bio-techno-social actors, in which the anthropological centrality gets inevitably contested, if not made irrelevant. Humanity will be living (in) dynamic, "interactive milieux" in which the tech–human dualism will be resolved as a multi(f)actorial coalescence that will define its own horizon of existence and action at all times. Building on Jacob von Uexküll's ([1934] 2010) notion of *Umwelt* (translatable as "environment-world"), Rowan Bailey and colleagues (2019, 9) write that "this description of an ecology . . . opens itself out to the entanglements and shifting perspectives within a play of relations." Most important, in this play of relations, technology has a crucial role on/in the coming into being of the milieu: "environing," Sverker Sörling and Nina Wormbs (2018, 103) argue, referring to the gerund of the noun, "consists of processes whereby environments appear as historical products, and technologies as the tools required for the environing to take place." The unique aspect, when it comes to symbiotic autonomous systems, is that they have two characterizing features: increasing autonomy and potential self-evolution.

For the first time in human history, a whole new set of non-fully organic systems are becoming able to autonomously shape and refine their own interfacing with/in the world, adapting to and getting shaped by it. Personal and city digital twins will *really* become living entangled systems able to self-organize their own boundaries while interacting among them, as well as with their constitutive parts.

Within this scenario, humans are dethroned for good from their self-imposed (literacy-based) role as masters of the world, rather requiring them to become coactors. And yet they will also profit from the spillover of such symbiotic relations, by seeing physical as well as cognitive facilities empowered. Humans too are in their (re)making through technology; they actually have always been in this kind of performative hybridization.[12]

                              *   *   *

What does this mean, concretely? Here we advance a geopolitical reflection that will facilitate the link to chapter 4. In recent decades, global trade has become an increasingly complex affair manifested by the emergence of transnational multistakeholder relations that cut across and remold nation-states' sovereign legitimacy. This is just another sign of the sociotechnical symbiosis discussed above under a geopolitical guise. In this scenario, indeed, ICTs play a major role: research has shown the "misalignment" between the internet as a commons infrastructure and the legitimacy of sovereign powers (Mueller 2019). Traditional categories such as "national" and "international," as well as "market" and "state," are no longer sufficient to account for today's tech-based geopolitics. For instance, Yu Hong and Thomas Goodnight (2022) argue, with specific regard to China, that "China's so-called Intranet also reveals entanglements with foreign capital, foreign technology, foreign markets, and foreign labor." This means that ICTs have impacted on nation-state sovereignty in multifarious ways, reshaping established power relations among and within states, fostering new public–private alliances across borders, as well as creating the preconditions for new asymmetries at glocal and *collectual* (individual + collective) scales. In this context, the risk is that, through ICTs, cyber-geopolitics pairs with forms of exploitation, especially connecting developed and less developed regions of the world. Linnet Taylor and Dennis Broeders (2015) denounced this with regard to Africa, where foreign actors' eagerness to control the deluge of ICT-derived data that come from the continent is alarming. Most significant—although hardly surprising—is the lack of agency of Africa, African institutions, and African people, who are chiefly considered recipients of top-down ICT-related investments and policies developed out of the continent. While calls for indigenous data and technologies are voiced (Mutsvairo and Ragnedda 2019), still at the end of the millennium, Manuel Castells (1998, 162) considered the whole African continent a "black hole of informational capitalism," by which he meant a geopolitical region that was out of the datafield. And yet, over the past three decades, African countries have witnessed a boom of ICTs, leading their economies to be immersed into the datafield and be part of a (global) geopolitical struggle.

It is no surprise that, in this scenario, the concept of digital sovereignty has become a buzzword. At the heart of the matter is control over data and tech infrastructures. At the same time, as scholars (Glasze et al., 2023) point out, technology governance becomes part and parcel of geopolitics when power relations heavily influence how a technology is developed, implemented, controlled, and used. From this perspective, digital sovereignty can be best regarded as a macro entangled cyber-geopolitical dimension that contests and resists linear (agent–structure) readings on which nation-state sovereignty rests. It is a whole new dimension that comes into being.

It suffices to mention the *Belt and Road Initiative* (BRI), a major pluri-decennial plan of investments connecting, by land and sea—the New Silk Road and the New Maritime Silk Road—China to Europe, while also involving Africa. In this scenario, a recent report of the Forum on China-Africa Cooperation (FOCAC 2019–2021), which outlines profound tech-based synergies between China and African countries, states that "in the future the application of quantum physics principles on computing will have huge implications. China will leverage its strengths in these sectors to support African countries to the best of its ability." Under the guises of "support" and "tech for good," it is likely that major quantum actors—not only Chinese—will stretch their influence on third parties and/or test their technologies in less technologically advanced countries.

To be sure, this kind of power asymmetries affects both North and South of the world—if this distinction is still of any value. In 2019, the European Commission (2019b) warned against the "digital dependency on non-European providers and the lack of a well-performing cloud infrastructure respecting European norms and values." One year later, the new president of the European Commission, Ursula von der Leyen (2020), spoke of the EU's digital sovereignty in these terms: "to make its own choices, based on its own values, respecting its own rules." Von der Leyen's target were those digital services and tech capacities—often offered by private multinationals to which also governmental bodies and public actors resort—whose uncontrolled flourishing would provide the basis of a new form of data-related colonialism.

In this sense, the other conceptual cornerstone is that of technological self-determination, also echoed by von der Leyen's words. But what *kind* of self is at stake? Certainly, not only the self of the single person, which, as seen, gets literally doubled by and through tech devices and drowned into the datafield. More specifically, the self at stake here is a fractal concept that cannot be reduced to a static mapping of actors and relations. This is so because it is the fundamental attributes of identity (individual *and* collective) that the digital contributes to remix. At stake, then, is the need to design and operationalize a kind of collectual self-determination that moves away from prioritizing either certain actors—private or state actors—or values—oftentimes economic competitiveness over social inclusiveness or environmental sustainability—to rather enact a balanced ecosystem as a whole and over time. In this regard, for the EU (as well as for other regions, such as the African digital single market), the establishment of a legally binding—still missing—*digital polity* becomes key to foster a full-fledged digital sovereignty that is able to protect people as *both* individuals and collectives (Calzati 2023).

How to Move On?

As soon as one thinks, perceives, and designs something through any *dispositif* and operating system, it acts on it (and is being acted upon). As Maturana and Varela (1987 242) write, "By existing, we generate cognitive 'blind spots' that can be cleared only through generating new blind spots in another domain." Data models and languages are complementary (embodied) *dispositifs*: they are epistemological triggers and ontological framers; they manipulate the world in idiosyncratic ways as much as they define their own horizon of truth. This brings about a sense of responsibility concerning conscious life (i.e., to know to be living). On this, the same authors further warn that "*the knowledge of knowledge compels*. It compels us to adopt an attitude of permanent vigilance against the temptation of certainty"

(245). To keep this in mind is more than a duty: it is a call for ecological action. To disregard this, instead, is a call for fiction, if not self-destruction.

In a general sense, digital entanglement is a Gordian knot of parameters. Its correlative is complexity. Looking back, benefiting from hindsight, we might say that big data have been an intermediary stage between classical and quantum technologies. Just as synaptic connections in the brain at rest, data exist as almost inert waves in the datafield unless and until a question, a query, is asked. The upcoming turn is "qualitative big data," that is, fast and actionable data, which will help harness the data to make the right decisions at the right time. With generative AI delegating the social control of language as a new application of the digital operating system, complexity will be instantly increased by an unfolding dual nature of language, human and/or machine driven, that will require a new operating system that disambiguates useful meaning from sheer performativity.

By collapsing the explicable cause–effect link and dispensing with the persisting connectivity at the base of data-driven technologies, *datacracy* has sublimated the binary logic of its own ecology. This requires, in turn, a new vocabulary and new tools for making sense of the new scenario. To bring quantum physics into the picture forces to rethink reality from scratch, insofar as the emerging actors, factors, and discourses of the quantum ecology (chapter 4) will interlace with existing ones and bring about an onto-epistemological shift (chapter 5).

4 Politics: The Realpolitik of Quantum Fields

The technology that is now developing and that will dominate the next decades is in conflict with traditional, and, in the main, momentarily still valid, geographical and political units and concepts. This is a maturing crisis of technology.
—John von Neumann ([1955] 1986)

How can we simulate quantum mechanics? Can you do it with a new kind of computer—a quantum computer? It is not a Turing machine, but a machine of a different kind.
—Richard Feynman (1982)

When John von Neumann uttered the words in the first excerpt, quantum technologies were nowhere in sight. What makes these words extremely relevant is von Neumann's clairvoyance of the disrupting force of technological innovation and—as the H-bomb had already demonstrated—its innervation with power and political dominion. Today, we see history repeating itself: von Neumann's words map onto Richard Feynman's, who preconized the advent of quantum computers as a "different kind" of machines. But different, *how*?

From today's perspective, with data-driven technologies well established in everyday life and quantum computers in sight, it is safe to consider the digital transformation as a transition phase that will blend with the emerging quantum ecology, as soon as the actors, factors, and discourses of the latter will consolidate. However, while the overlap between the language and digital ecologies provoked an epistemological crisis, the overlap between the digital and quantum ecologies (and also the language one) will likely produce an *onto*-epistemological crisis. This is so because the physics' tenets at the basis of quantum *information* technologies (QITs) are essentially different from the classical ones that frame both the language and data-driven technologies. The shift will not be sudden, but it will be profound and long-lasting.

In this context, both QITs (e.g., quantum computers, quantum internet) and quantum actors (e.g., the EU, the US, and China, as well as companies) can be regarded as emerging quantum instantiations whose actions and interactions inevitably shape the new sociotechnical field. In this respect, the configuration of the quantum ecology will depend on how QITs and quantum actors will evolve, as well as the narratives they will be able to build around them(selves).

To retrace the genesis of the quantum ecology, it is useful first to understand how major actors harness and tackle current prospects and challenges of technological innovation, especially AI. Then, we will discuss major trends and discourses in QITs, whose development will have socio-political-psychological-ethical effects.

Prospects and Challenges of AI

While apparently on different trajectories, scales, and timelines of evolution, the link between AI and QITs is deeper than it appears. Notably the prospects and future developments of the former can be capitalized by the latter toward what could likely be a future merging of the two. By drawing a felicitous distinction between *complex* (computational) and *difficult* (skill-related) problems, Floridi (2019, 11) writes that "the future of successful AI probably lies not only in increasingly hybrid or synthetic data, but also in translating difficult tasks into complex tasks. How is this translation achieved? By transforming the environment within which AI operates into an AI-friendly environment." To do this, the environment needs to be "enveloped" through increasing datafication and computation-oriented design of physical behaviors and phenomena. This is, indeed, what QITs might favor, by enhancing and speeding up the realm of computability: researchers (Bibri 2018; Hoofnagle and Garfinkel 2022) envision not only the central role that quantum computers will play in such a process, but also that mass-market commercialization of quantum computing will likely be (at least initially) in the form of cloud services, thus facilitating a hybridization with emerging AIs, also in the form of cloud services.

At the moment, however, AI—in its multifarious experimentations—still faces some structural challenges. As Kai-Fu Lee (2018, 14) notes, "Today, successful AI needs three things: big data, computing power and the work of strong algorithm engineers." True, the scenario we described in chapter 3 produces an ever-increasing amount of data upon which data regimes are willing to capitalize, but this requires ever more crunching power and skills, as well as storage.

More Data and More Computing Power

Not only today does the data-based "grab-all logic" run at full steam, but it has become increasingly affordable. Humanity entered the so-called zettabyte era when, in 2016, the global IP traffic first exceeded one zettabyte (10^{21} bytes), and it continues to grow staggeringly (in this respect, the often-overlooked aspect is the carbon footprint of such growth; cf. Hasselbalch 2022). Alongside, the costs of data storage have decreased thanks to the consolidation of "cloud," "edge," and "fog" computing.[1] These technologies are at the core of data regimes' materialization on the geopolitical global stage, that is, how they play out their mutual relations, their conflicting interests, and their strategies at various national and international scales. Technology *is* about enacting fields of actions and influences, as von Neumann knew only too well.

The need for literally creating new space for the storing of data is unstoppable. The alternative would be to either dispose of such data or get engulfed by them, thus augmenting latency (and slowing down services) and/or making it impossible to take proactive actions by means of data—a failure that data regimes cannot afford. The need for data space is so alluring that data centers now also come in the form of service offerings. US and Chinese tech giants are on a rush for building new hyperscale data centers on their territories or—more problematically—dislocated around the world.

For instance, Google opened new data centers in northern Virginia and Tennessee in 2019 and initiated discussions for other sites in Texas, Ohio, Nevada, Nebraska, and Arizona, with centers already present, among other areas, in Europe and South America. Apple, Amazon, Meta, Microsoft, and IBM are on the same page, spending billions each quarter for building and equipping massive storing facilities in the US and abroad. As for China, in 2019, the investment firm Bain Capital favored the merger between the

Singapore-based Bridge Data Centres and the Chinese data center provider Chindata, with the goal to establish a pan-Asian data center giant.

Of course, increasingly powerful data centers require increasingly efficient infrastructures—let alone ever-efficient energy production systems—in order to be maintained. In 2019, for the first time in history, companies spent, in a year, more money on cloud infrastructure services than on data center hardware and software. More important, it is the openness of the architectural framework that supports these solutions to make the difference: indeed, trends show that storage and computable capacities are moving toward forms of "co-location" and "hyperconvergence," which help maximize resources usage and energy efficiency. By the end of 2022, 75 percent of enterprise-generated data were created and processed outside of traditional data centers or even the cloud.

Making Data Actionable

Alongside the increase of data and computer power, the issue of the searchability/manageability of data is key. Even capillary surveillance such as that put forth by the NSA has not been able to prevent terrorist attacks (e.g., the Boston Marathon) or mass shootings, especially when these acts came from isolated wolves. *Datacracy*, as seen, is not (yet) a totalitarian system: it prefigures and configures probabilistic scenarios, largely capitalizing on vast arrays of databases that, however, are not easily scrapable. Already in 2016, William Binney, a former NSA official then turned into a whistleblower, stated clearly that the surveillance program enforced by the NSA had turned into a "bulk data failure" due to the expensive and ineffective amount of information collected.[2] In other words, the surveillance program has (had?) too much "raw" potential that could not (yet) be efficiently exploited. For better or worse, solutions to this issue are in sight.

At stake is the quest for more efficient ways to acquire/collect data, preserve/circulate them, clean/synthesize them, and process/retrieve them. Data are expected to fit different purposes, by different actors, in ever more entangled ways. Hence, they need to become easily *actionable* upon request in increasingly faster ways.

Lee (2018) rightly argues that deep learning (and now self-supervised learning, which is a form of machine-autonomous deep learning where data are automatically labeled by leveraging the relations between different inputs) has been the technological breakthrough that turned the IoT from a futuristic scenario into a realistic one. And deep learning is still the innovation whose long tail will continue to have a major impact in the years to come. In fact, deep learning can be harnessed for many purposes. Notably, the refinement of how data are made accessible and pertinently processable has gone in this direction.

In order to make the IoT more and more efficient and reliable, tech innovation is pointing toward deep learning–based solutions for the processing of real-time data capitalizing upon distributed learning paradigms. Deep learning, then, is not only the breakthrough by which data-driven technologies have become increasingly adaptive and autonomous; it is also the means to leverage on the increasing amount of data produced every day, so that it is put to use instead of vegetating in databases. AI functioning will depend more and more on . . . AI. From here, the advent of context-sensitive algorithms that will be able to autonomously decide what is the relevant and pertinent data to be accessed, processed, and circulated in any given situation is not hard to predict.

For an example that fits the present discussion, in a recent article (Jindal et al. 2020), researchers have presented a tensor-based big data management technique to reduce the dimensionality of data gathered from the internet-of-energy (IoE) environment of a smart city. Resorting to a series of tensor operations, the idea is to reduce the dimensionality of the huge amount of data gathered by smart city's diffused sensors. The goal is to obtain a core set of data to be stored in the cloud in a reduced form, as a sort of "simplified" version of the richness of data acquired. This makes data easier to handle and process when summoned up for any purpose. Using an (approximated) analogy, the proposed solution might be seen as a mix between zipping a traditional image file and saving it in a less heavy format (say, "jpeg" instead of "png").

Interestingly, tensor networks are increasingly adopted also in quantum physics, especially for studying the nature of space and time, but also (and this is most important here) in the context of the development of quantum computing. Simply put, a tensor is an algebraic object that describes multiple values at the same time or, better, the relation between them. Hence, tensors are a sort of mathematical shortcut that makes it easier to handle equations. As a step further, the networking of tensors is able to account for, and make mathematically operable, complex systems such as a quantum circuit, which can be understood as a product of linear operators on various quantum states. Tensor networks, therefore, help not only to mathematically "visualize" a quantum circuit but also to mathematically work with it. Before discussing QITs, let's map the current landscape of existing quantum sensing technologies.

Quantum Sensing Technologies Already in the Air

Today, the term *quantum technologies* refers to a varied set of technological innovations, some of which are already in use. As physicist Rainer Blatt (2020, 1) writes, "The powerful quantum framework led to revolutionary technical advances for a century, such as semiconductor technology and the entire electronics industry based on it, enabling computers and the information age." This means that devices such as smartphones, LED screens, and even the PCs we are using to write at this moment are based on the quantum mechanics formalized at the beginning of twentieth century. Our life is flooded by quantum-based technologies; quantum onto-epistemology has already put its seed in the present sociotechnical condition, and it cannot but grow in the near future.

Broadly speaking, it is possible to distinguish between "quantum sensing technologies" (e.g., quantum metrology, quantum navigation, and quantum imaging) and "quantum information technologies" (QITs) (e.g., quantum computation, quantum simulation, and quantum communication). While the latter are still in the developing phase, the former already innervate our everyday life.

An example of quantum sensing technology is magnetic resonance imaging (MRI), now widely used in many contexts, especially for medical diagnosis, which exploits the spin of the proton contained in the hydrogen's nucleus by applying a magnetic field to it that makes the spin flip. Sensors then keep track of such flips to re-create an image of the body. A further example is the atomic clock, which measures time by counting the oscillations of electrons when they are excited (i.e., receive or lose energy). The best atomic clocks are so precise that they will not miss a second in the whole age of the universe. Atomic clocks are used for synchronizing time across the globe and to make GPS extremely precise, by aligning data coming from orbiting satellites.

These are just two examples of how quantum physics has already entered our life through technology. This materialization, in turn, disrupts and reworks people's way of being and acting in the world. As seen in chapter 3, ever-localization is one of the preconditions of *datacracy* as a de-subjectified system in which individuals are turned into data, time is kairological, and space is reduced to point-like granularity. GPS makes both distance—the geometry of space—and orientation—its relativization—irrelevant. This imposes a flat transparency on the subject that leaves no room for qualitative connotation: the subject's point-of-being in the world is shrunken to a mere set of entries. Similarly, time coalesces and sublimates into what one can observe at every moment on the map. This idea of time remolds experience and configures a condition of inescapability that annihilates self-determination intended as a projection into the future and a recollection of the past. To be is no longer to become, but to leave traces and be traced—as observables—within an ever-reconfiguring dimension. This dimension has the features of the myth: time and space are no longer phenomenological coordinates, but tech-transcendental features. As a consequence, existence gets metamorphosed into a proxy in the realm of probability. In doing so, however, what is being enacted through technology is an ever-possible-subject to whom, eventually, to be denied is the very possibility of not-being (traced).

Quantum Information Technologies

> The fundamental laws necessary for the mathematical treatment of a large part of physics and the whole of chemistry are thus completely known, and the difficulty lies only in the fact that application of these laws leads to equations that are too complex to be solved. (Paul Dirac 1929)

Dirac's point might be a bit closer to be solved. Today, indeed, humanity is entering what Dowling and Milburn (2003) have called the "second quantum revolution": after the first theoretical quantum breakthrough at the turn of the twentieth century, now quantum physics' behaviors are being systematically applied to QITs. Notably, this class of technological innovations relies on a different approach to computation and information processing. This is what Feynman envisioned when speaking of quantum computers as different in species from traditional ones. This diversity lies in the possibility of making use of phenomena such as superposition and entanglement, which are fundamentally probabilistic, thus reconfiguring the scope of the "possible" once such phenomena are engineered in technology. QITs perform computation by manipulating particles, not classical bits: as soon as quantum phenomena breach into applied fields and are incorporated into macro technological apparatuses, then the latter bear the mark of the former in their effects (in the same fashion as the internet *did* wire the world).

Under the umbrella term of QITs falls different research streams and technologies that are worth disentangling. On the one hand, there is *quantum computing*, which encompasses quantum simulation and quantum computers; on the other hand, there is *quantum communication*, within which one can distinguish applications such as quantum teleportation, quantum networks, and quantum internet. Interestingly enough, these are parallel streams of research and application, meaning that a running and working quantum network—and a broader quantum internet—are independent from the development of large-scale quantum computers. This is because a quantum internet can in principle already emerge by connecting very simple (fault-tolerant) quantum devices, while a proper quantum computer requires the aggregation of large amounts of quantum bits (qubits).

Quantum Internet

Several prototypes of quantum networks have already been implemented around the world. These are sometimes referred to as prequantum networks because they still rely on classical nodes at the ends of the network. While full-fledged quantum computing networks are still far down the road, as they depend on the development of stable quantum computers, today one promising avenue to build quantum networks is that of relying on ground quantum repeaters. As the term suggests, these repeaters allow for the encrypted distribution of entangled keys to be passed along the network; alternatively, research is ongoing to enable entangled particles to be temporarily stored along the network and retrieved when needed. Quantum ground repeaters are, for instance, the pillars of the Dutch quantum network developed at the Delft University of Technology in the Netherlands. A different path is the one pursued by China, working on a satellite-based quantum network. In 2016, a Chinese research team led by Jian-Wei Pan, sometimes referred to as the Chinese father of QITs, launched the world's first quantum satellite, which in 2017 made possible a video call between Beijing and Vienna using quantum encryption. This meant that no one with whatever existing technology could actually tap into it, or the communication would break down. In June 2020, Pan and his team improved the satellite quantum communication and succeeded in transmitting keys for encrypted messages between the satellite and two ground stations far apart. With the consolidation of ever-wider quantum networks, the envisioned outcome is to federate these into one or multiple quantum internet(s).

Quantum Computing

As Eline de Jong (2022, 14) writes, "In the near term, quantum simulation is the most interesting form of quantum computing." Quantum simulators are software programs run on classical computers to simulate quantum operations (although, at the cost of intense crunching). Quantum simulators are useful to target the study of quantum systems and specific physics problems, or also to create better algorithms, but they can at best assist quantum computers—not replace them—as these latter are more complex programmable (hardware) machines that can natively perform quantum operations.

At the end of 2019, Google claimed to have reached, through its Sycamore computer, what for decades has been the holy grail of computer engineering: "quantum supremacy" (Preskill 2012). This means that Google's quantum computer was allegedly able to make a calculation much faster than the fastest supercomputer currently available (which runs according to classical physics principles). According to Aaronson, who reviewed Google's paper for *Nature* before publication and is one of the most recognized authorities in the field, quantum computers are expected "to change the fundamental scaling behavior of algorithms, making certain tasks feasible that had previously been exponentially hard."[3]

But how does a quantum computer work? While a traditional computer works according to a binary logic of 0s and 1s, processing these bits of information sequentially, in a quantum computer, bits—called *qubits*—are superposed, meaning that they can assume, *simultaneously*, two states—0 *and* 1—or, more correctly, an amplitude for being 0 and an amplitude for being 1. To do that, it is necessary that the chip at the core of quantum computer runs in particularly delicate conditions, close to absolute zero (–273.15°C) and in complete isolation, so as not to let the system decohere, which would make the qubit superposition disappear. Harder still is to "assemble" qubits together, making interactions among their amplitudes possible and generating entangled states within the system. Google's

Sycamore has fifty-three qubits, which means 2^{53} amplitudes—potential states of qubits—or about nine quadrillion.

What has the Sycamore computer achieved? Google asked its computer to make a calculation complex enough to "occupy" close to the nine quadrillion options available. The operation took Sycamore three minutes and twenty seconds, while it is claimed that it would have taken 10,000 years to 100,000 traditional supercomputers to do that (although this was later criticized by IBM).[4] It is a milestone, but not only that. It is a turning point: it suffices to think that still in the 1990s, some computer scientists deemed the construction of a quantum computer impossible. The next step, as Aaronson suggests, is "quantum error correction: a technology that, in theory, should be able to keep qubits coherent for vastly longer amounts of time by cleverly encoding them across many physical qubits."[5] Quantum error correction is the attempt to encode logical qubits, which can then arbitrarily correct their errors and remain coherent for longer periods (Georgescu 2020). At present, it is still difficult to maintain a balance between the coherence of the whole system and the adding of qubits, since this process needs to be done in array, thus requiring the reconfiguration of the whole system to keep it coherent.

Recently, however, researchers from the Tokyo University of Science and the Sydney University of Technology (Mukai et al. 2020) have introduced a novel two-dimensional design that can contribute to tackle this problem. This novel design keeps all qubits on the edge of the system, simplifying the wiring that keeps the system coherent, without recurring to three-dimensional architectures that have made so far the scalability of qubits difficult. Another line of research (Campagne-Ibarcq et al. 2020) builds on the pioneering work of Daniel Gottesman, Alexei Kitaev, and John Preskill (2001) to create a protocol that is able to almost simultaneously detect the noise to which qubits are subjected and "canceling it out" through an oscillator. In this way, the coherence among the qubits is maintained longer. A third option has been envisioned by a team from the University of Sydney in partnership with Microsoft (Pauka et al. 2021): they made it possible to control and manipulate qubits through a new chip that works at the same close-to-absolute-zero temperature as the qubits. This entails a huge simplification in the wiring of the quantum computer, thus opening the way to easier scalability. More recently, by using a fault-tolerant surface code, in February 2023, the team at Google Quantum AI (2023) initiative showed that quantum error rates can be actually reduced by increasing the number of qubits used to perform the error correction. At the end of the same year, just a few days apart from each other, a team led by Harvard University (Bluvstein et al. 2023) and a team led by IBM (IBM Newsroom 2023) have announced, respectively, the first programmable, logical quantum processor, capable of "early error-corrected computation" and a new way of modularly scaling quantum chips up to quantum systems. While new announcements go by the day based on different research paths, quantum error correction and qubits scalability remain the core issues to tackle for the development of stable quantum computers and (possibly) their marketization.

Potentialities and Limits of Quantum Computing

Nowadays, quantum computing is often thought of and described as a way to process things faster compared to what traditional supercomputers can do. This is certainly true; however, as soon as such an increase of speed becomes exponential, a substantially different dimension reveals itself, one in which quantitative acceleration generates a qualitative leap. In other words, quantum computers could make the abovementioned technological "envelopment" of {reality} affordable to such an extent as to radically transform the very nature of what is assumed to be real. Dirac's admonition might (relatively) soon be overcome.

On the one hand, it is important to avoid technological hype and mystification. Views on the front of what quantum computers can do remain cautious. In 2008, Aaronson wrote an interesting article on "The Limits of Quantum Computers"[6]:

> According to our current understanding, [quantum computers] would provide dramatic speedups for a few specific problems—such as breaking the cryptographic codes that are widely used for monetary transactions on the Internet. For other problems, however—such as playing chess, scheduling airline flights and proving theorems—evidence now strongly suggests that quantum computers would suffer from many of the same algorithmic limitations as today's classical computers.

Notwithstanding the limits that quantum computers will keep facing in the future (and bearing in mind that, since Aaronson's statement, AI has massively progressed beyond expectations), the point we want to make is simple: even though quantum computing will not solve all computational problems, the technological horizon of quantum computers is constantly changing and pushed, every time, a bit farther. Recently, for instance, significant steps have been made in the understanding of what quantum computers can do *uniquely* in comparison with traditional supercomputers, entailing not only a computational acceleration but a qualitative leap from today's scenario.[7] This does not mean to fall for quantum wizardry, but to make room for the very possibility that a radical sociotechnical shift is on its way. Indeed, quantum computers are *systemic technologies* as the steam engine, electricity, and the internet have been (WRR 2021). As such, they will innervate into the social and codevelop with it, bringing deep unpredictable effects at (geo)political, cultural, psychological, and ethical levels. As de Jong (2022, 9) nicely put it, the fact that commercial quantum computers are not here yet "should not rule out the possibility nor the responsibility to anticipate that future. Not only *despite* but also *because* of the early phase of development and application, we should start thinking about the ethical, legal, social and policy implications of quantum technology." If futuristic predictions are never a safe ground, historical hindsight can help understand the magnitude of the path that has just opened. When the technology at the basis of the internet was developed by the US military in the 1970s, the Web was not even close to being formalized. In the span of fifty years, virtually the whole world has been connected with the unimagined consequences we discussed in the previous chapters. Currently, quantum computers and AI are developing fast and in unpredictable ways; what is likely to happen, though, is their convergence in the near future: in fact, the implementation of neural networks (Beer et al. 2020) and machine learning (Tacchino et al. 2019) on quantum computers will bear vast implications for research and innovation. In the long term, when the merging of AI with quantum machines will be consolidated, it might well represent an enabler for an extended, flexible, and possibly autonomous synthetic {reality}.

Today, humanity finds itself—once again—on the verge of a technological turning point where old and new ecologies overlap. The process might take some time to come to light, but it is only a matter of *when*, not *if*. Our goal here is to explore the "how" of such a process—how disruptive, how deep, how long-lasting—in order to unveil from the outset its key drivers, as well as its socio-economic-cultural potentials and drawbacks.

New Frontiers

Research and development on quantum computing will constitute a deeply disruptive game changer in a number of sectors. Quantum computing will support not only theoretical fields, such as math and physics, allowing for the simulation of complex systems, but also practical fields, such as chemistry and biology, cybersecurity, communication systems, military, astronomy, and computer sciences. Once introduced, quantum computing will boost research and development, enticing exponential advancements that today one can barely foresee. No field will be spared.

Chemistry, as hinted by Paul Dirac himself, will likely be among the first fields to be impacted by quantum computing, since it is a field based on a high degree of formalization that lends itself to be processed. Today's classical computers used in the simulation of chemical processes can only reach that far, in that some of these processes are too complex to be treated classically. Hence, the possibility to use quantum computers to simulate these processes will very likely usher chemistry into a new unmapped era of development. In 2023, quantum super chemistry—a kind of chemistry performed at very low temperatures in which atoms or molecules in the same quantum state collectively react more rapidly than do atoms or molecules individually— was observed for the first time. The possibility to subsequently model these super chemical reactions might be supported by quantum computers.

Consequently, to be deeply impacted will also be molecular biology. Here, the possibility of accurately simulating the combination of amino acids— the protein folding process at the base of life—will lead to new advancements with far-reaching implications in a number of sectors (e.g., artificial lifeforms). Already today, it has been shown how quantum computing can give an edge in the predictive understanding of the binding of gene regulatory proteins to the genome. In the future, the decoding of how proteins work will have concrete repercussions in medicine, leading to increasingly customized treatments for now incurable diseases. Scientists could literally engineer the coding system of life to "fix it" whenever it acts wrongly.

Another field that already today heavily relies on automated predictive systems is finance. With the introduction of quantum computing, the whole sector will be deeply disrupted, most likely requiring new ethical-legal frameworks. Applications of quantum computers to stock market and brokerage as well as to portfolio selection and risk simulation will lead to the creation of ever more accurate scenarios. Hence, it might be necessary to proactively regulate the sector for preventing an ill-driven abuse of quantum computing that might negatively impact large portions of the society, increasing social inequalities and widening economic divides.

Other highly automated sectors are logistics, manufacturing, and supply chains. Today's computers already make the efficient rationalization of these sectors possible, optimizing production and delivery scheduling, energy distribution and consumption, as well as refining interoperability across the actors along the chain. With the consolidation of the IoT, the increasing stockpiling of data might become impossible to manage for classical computers, and here is where the application of quantum computing will really represent a necessary leap, allowing for an integrated tackling of multiple issues at once.

Quantum computing will also be used endogenously, so to speak, to optimize the functioning of software, data centers, and data distribution networks. The need for designing and testing the working of codes, networks, and storage systems is expanding greatly and fast. With the development of sensing environments such as smart cities, it is critical to know not only how the hardware and software infrastructures behind these environments perform but also how to tackle possible malfunctioning. This is an area in which the computational power of quantum computers will turn out to be essential.

Last, cryptography. Currently, most encrypted transactions rely upon asymmetric cryptography, which provides different keys to code and decode a transaction. These keys are produced by an algorithm—developed by Peter Shor in 1994—on the basis of the factorization of huge numbers (every integer can be factored as a product of prime numbers). Although today it still takes a huge amount of time for a computer to discover these factored keys, quantum computers may easily breach current cryptography. According to the Canadian Global Risk Institute,[8] there is a "one in seven chance that some of the fundamental public-key cryptography tools upon which we rely today will be broken by 2026 and a 50% chance by 2031." This is an enduring problem due to the fact that, even if today quantum computers are not here yet, encrypted data—including sensitive data—can be archived and hacked later. Along this line, "quantum safe" and "postquantum" cryptography have already attracted interest as a way to make cryptography unbreachable also in the quantum era. In 2022, the US National Institute of Standard Technology opened a call for the design of quantum computer-proof algorithms, which led to the selection of four algorithms for establishing a first postquantum encryption standard.

Another avenue is "quantum cryptography," which builds upon quantum computing to create quantum cryptographic solutions.

Quantum Geopolitical Fields

> If new means for perception and surveillance are made available (to see new spaces, new scales, new traces, new crimes), then governance will conform itself to the vacuum opened up by new vision machines and to the demands of whatever is now available to observe and control. (Bratton 2016, 8)

In recent years, QITs have not only entered the political agendas of various countries but also gained a central stage, becoming the target of public–private investments and roadmaps for decades to come. This witnesses the extent to which political actors across the globe are now aware of the potentialities of quantum technologies for the twenty-first century. Among others, the EU, the US, and China all have released official documents that acknowledge the deep changes QITs will bring to societies.

Normatively, three major competing visions are usually identified when it come to the steering of tech innovation (Schneider 2020): (1) a *corporate-driven* approach (e.g., United States), based on market deregulation and favoring economic competitiveness among tech stakeholders and platforms; (2) a *state-led* approach (e.g., China) depending on authority-defined plans and striving for global technoeconomic leadership in strategic sectors while maintaining state control over social and moral behaviors; and (3) a *citizen-centric* approach (e.g., European Union) aiming to achieve tech innovation by safeguarding human rights and balancing economic competitiveness with social inclusiveness, democratic participation, and environmental sustainability.

In the corporate-driven approach, the public sector tends to either play a facilitating role or become de facto the client/recipient of tech solutions developed, controlled, and owned by private corporations. In this scenario, the blossoming of big tech corporations "is seen as positive both for innovation and economic growth and hence is fostered," leading to "extremely high revenues [that] allow these companies also the power to lobby governments" (Schneider 2020). While favoring a competitive landscape where innovation and economic success go hand in hand, this approach also shows drawbacks, especially due to the lack of contextual adaptability of the developed technologies (Kummitha 2020), as well as their limited social inclusiveness (Kalpokas 2022).

By contrast, China's state-led approach is regarded as technological nationalism and is meant to foster economic-political gains (Jiang and Fu 2018). Within this approach, public authorities create new lanes for tech innovation via a top-down logic in which it is up to state authorities to broadly dictate the direction to follow to both public and private actors. Heavy public backing through financial facilitations favors the achievement of mid- to long-term targets, all part of China's goal to reach technological market supremacy by 2025. Yet, this approach too presents shortcomings. Bureaucracy tends to stiff innovation due, on the one hand, to the enduring "fragmentation of the state governance structure [and] the poor coordination within the bureaucracy" (Sun 2007) and, on the other hand, to a bottleneck disadvantaging innovation by small and medium enterprises compared to big private and state-owned firms (Fu et al. 2016). Despite having transitioned from a "copycat" model to an indigenous technological paradigm, China's state-funded digital transformation still fails to foster a strong link between industry and research because of lack of incentives to experiment out of identified paths (Han et al. 2019).

Concerning the EU, the idea to foster a citizen-centric tech innovation is reaffirmed in various pieces of legislation and policy-orienting documents part of its digital strategy, among which the recent Declaration on Digital Rights and Principles (European Commission 2022), where it is stated the objective to "promote a European way for the digital transition, putting people at the center." Yet, how to properly design such a citizen-centric approach, making sure that it strikes a balance among all its pillars (i.e., fundamental rights, economic competitiveness, social inclusion, environmental sustainability) is still uncertain. What is envisioned is, de facto, a Digital Single Market (European Commission 2021), which aims to establish a technically secure and legally compliant backbone for the economically profitable sharing of data, thus privileging (1) individuals (e.g., consumers and companies) over the (societal) collective(s), (2) private actors over the public sector and noninstitutional actors, and (3) technical over nontechnical (e.g., literacies, trust, governance) aspects (Calzati and van Loenen 2023b).

At stake is not an evaluation of which approach is "best"—they all come with pros and cons, each responding to specific value priorities, socioeconomic visions, technocultural fields, and ethical principles—but the extent to which they will be able to coexist or, by contrast, whether one will take the others over. Most important—and aligning to the discussion at the end of chapter 3—to talk here of the "US," "China," and the "EU" means to refer to tech regimes within a broader and very much contested field, connoted by transversal cooperation and competition cutting through national, supranational, and subnational borders, while involving a plethora of other public, semipublic, and private actors. The matter is not much one of technological advancement, but global supremacy. For instance, concerning specifically quantum technologies, within the EU, France and the Netherlands signed in 2021 a memorandum to boost synergies toward further research and development of quantum technologies; bridging the two sides of the Atlantic, the National Research Council of Canada and Germany are also collaborating in the same area. Countries have also established partnerships with companies and research centers: in 2019, Microsoft opened the Microsoft Quantum Lab on the Delft University of Technology campus; Google has a dedicated research center in Tel Aviv, while IBM has launched a quantum computer in Germany.

We certainly do not want to naively consider quantum technologies as the *causa prima* of a new geopolitical order; rather, we aim to discuss how quantum technologies can impact on governments and governmentality and, in turn, how they are shaped through policy guidelines. The best way to do this is by looking at the discourses that the EU, the US, and China have so far voiced. While the released documents do contribute to a form of quantum hype (Smith 2020), currently these remain the best resources we have to map emerging strategies. Moreover, these documents are good indicators of the flux of funding destined to new quantum technologies in the upcoming years—at least by public actors, which are forced to some form of public accountability (more in the West than in China)—thus zooming in on possible developments.

*   *   *

In October 2018, the European Commission launched the "Quantum Flagship,"[9] an advanced research and development initiative funded with at least one billion euros over ten years. The major goal is to keep the EU abreast of quantum tech innovation, fostering a competitive ecosystem made of public organizations, private actors, and academia. As a long-term objective, the initiative focuses on the emergence of a quantum networking infrastructure (or "quantum Web"), particularized as "quantum computers, simulators and sensors interconnected via quantum networks distributing information and quantum resources such as coherence and entanglement." At the occasion of the launch of the Quantum Flagship, Roberto Viola, director general of DG Connect, recognized "quantum [as] a highly strategic area for Europe. We must master it, both to deliver life-changing benefits for our citizens in fields like health, energy and cybersecurity, and to secure our technological sovereignty in a competitive field."[10]

In March 2020, the EU "Strategic Research Agenda"[11] of the Quantum Flagship was released. The document identifies four major areas of research and development of quantum technologies: (1) communication, (2) computing, (3) simulation, and (4) sensing and metrology. These areas, which are expanded in detail in the document, can be loosely mapped onto the fields of short-term and mid-term applications of quantum technologies identified in the previous sections.

From a political point of view, it is worth noting the EU's acknowledgment that quantum technologies will clearly pose an issue of sovereignty, as they will constitute the "critical building block for the future economic development and digital self-determination of societies."[12] In fact, to be acknowledged is not only the revolutionizing impact that quantum technologies will have in several fields but also the deeper sociocultural implications of such impact, which will question the effective self-determination of societies.

The document also clearly positions the EU in contrast with its major competitors—the US and China—and recognizes the need to incentivize synergies within the EU to fill financial and know-how gaps: "In contrast to the US risk capital approach and China's state-capitalism system," it is stated in the document, "Europe has not yet found its own way to meet this challenge."[13] One step has been the launch of the European Quantum Industry Consortium (EQIC) in April 2021. The EU not only denounces the risk of lagging behind, but it is especially aware of the limited availability of quantum-savvy scientists and engineers across member states. Hence, the Agenda sets as a priority "to significantly increase the number of trainees in this sector in order to meet the foreseeable demand,"[14] also through international partnerships.

*   *   *

In September 2018, the US subcommittee on quantum information sciences (QIS), part of the National Science and Technology Office (NSTO), published a document titled "National Strategic Overview on Quantum Information Science."[15] The document formalizes the strategic importance of QIS, defined as "a nascent pillar of the American research and development enterprise." The document provides general guidelines concerning the need for coordination across national agencies, public and private institutions, and international partners, as well as the funding of an academic-led quantum-smart workforce. The standpoint is to maintain "a culture of discovery" able to harness the economic impact of QIS, which at present is still "uncertain," as well as guarantee national security. Such discourse aligns well with the normative US conception of research and innovation as forces to be best supported via heavy funding, competition and convergence of multiple stakeholders, and little regulation. The stress is put equally on national interests and international cooperation, conscious of the fact that the US does not have the power alone to systematically approach the development of quantum technologies.

At the end of 2018, the US administration concretized such an overview by turning it into the "National Quantum Initiative Act." The act jump-started research and development on QIS for the next decade, allocating a budget of $3.75 billion dollars. Furthermore, the National Quantum Coordination Office (NQCO) was established, the function of which is to coordinate all federal actors involved in QIS and to identify the "grand challenges" of QIS (i.e., "problems whose solutions enable transformative scientific and industrial progress").[16] This funding trend was also confirmed—indeed, strengthened—for the 2021 budget, which increased the aggregate federal funding by more than 50 percent, on the way to double it by 2022. In other words, the US administration has made QIS a top research and development priority together with AI, rivaling China's long-lasting involvement in the field. As Jon Lindsay wrote in *The Washington Post*, aligning to other analysts, "U.S. government investment in science and technology has long been the foundation of America's economic and military might. Yet competition with China is what really matters in this case."[17]

In February 2020, the newly established NQCO released a document titled "Strategic Vision for America's Quantum Networks,"[18] in which major priorities and goals of QIS are detailed. Among others: "national and financial security, patient privacy, drug discovery, and the design and manufacturing of new materials." Similarly to the EU, the stress is on "quantum networking" or "quantum internet," namely, "platforms that reliably link together quantum devices to develop applications that leverage quantum-enabled security, sensing, and computation modalities."[19] Here, the time span exceeds the ten-year frame, identifying mid-term priorities for the period 2020–2025, as well as long-term objectives for the next twenty years, when "quantum Internet links will leverage networked quantum devices to enable new capabilities not possible with classical technology."[20]

*   *   *

China identified quantum informatics as a key area in its thirteenth five-year plan (2016–2020)[21] for economic and social development. This document states that "we will develop quantum communication and a safe and ubiquitous Internet of Things and accelerate the development of synthetic biology and regenerative medical techniques."[22] The decade of advantage that China, according to international observers, has over the US and the EU (at least on paper) is the consequence of Chinese analysts' long-lasting belief that gaining an edge on quantum technologies will help stir the future of international politics to China's own benefit.

The declared goal of China is to become a technological superpower in the next few decades, and quantum technologies play a crucial role in this plan. Beyond government funding—which is not possible to quantify with accuracy: estimates are around $10 to $15 billion over ten years—private Chinese companies are also pouring investments into quantum research and development. In 2017, for instance, Alibaba announced a $15 billion plan of investments into its so-called DAMO Academy—"Discovery, Adventure, Momentum, and Outlook"—which aims to advance "foundational and disruptive technology," among which are AI and quantum computing. The ambitious overarching goal is to get two billion customers and create one hundred million jobs by 2036.

All this attention toward quantum research and development is already reflected in statistics. In 2017, Chinese organizations filed nearly twice as many patents related to QITs as the US, and more than 70 percent of academic QIT patents since 2012 have been awarded to Chinese universities, with US institutions a distant second at 12 percent.

China has worked for years especially on short- and long-term applications concerning communication systems and the delivery of unbreachable encrypted information across networks. While Google might have reached a major breakthrough with its Sycamore computer, this achievement risks to metamorphose into a reversed "Sputnik event" if not supported by continuous research, insofar as China has in the meantime invested heavily in research whose more immediate applications might turn out to be a better leverage for getting abreast of competitors.

*   *   *

Apart from the magnitude of investments, it is possible to detect various similarities between the EU and the US approaches to quantum research and development. First of all, the timing: both actors launched their quantum-related initiatives at about the same time and based their roadmaps over a ten-year time span (although US documents point out the importance of thinking already beyond 2030). Second, there is a shared focus on the development of quantum networks, which are expected to constitute the canvas of all subsequent applications. Third, both actors reassert the need for international cooperation, welcoming cross-fertilization and synergies. In this respect, both the US and EU are aware of the strategic necessity to foster and enlarge the cohort of quantum-savvy scientists working today on quantum research, insofar as to get an edge in the field will depend much on the possibility to attract, retain, and form a wide basin of experts.

By contrast, from what emerges from the official documents, China tends to have a more isolated approach. Witness funding Chinese scientists to return from abroad (Cao et al. 2020), although—as seen in chapter 3—partnerships with African (and South American) countries are also on the agenda. This approach aligns with "Made in China 2025," a document released by the Chinese government in 2015, setting the goal to turn China into a technologically independent power within a decade. This plan, which has been partially revised, has spurred criticism especially among Western trading partners, who claim that the state-supported plan breaches free trade agreements.

Beyond that, China's authorities provide precious little data concerning investments in quantum research. To this, private investments must be added (see, for instance, Alibaba), not rarely facilitated by political decisions and fiscal facilitations, which, however, might be subjected to changes over the years. Moreover, little insight is given on the detailed plans of development of quantum technologies, possibly because they are framed as a matter of civic–military national security. Hence, the documents depict a very broad and general picture, one indeed open to various interpretations. This communicative austerity is typical of Chinese authorities, who tend

to provide general guidelines for various sectors, listing some precise economic goals and more abstract conceptualizations of such goals (Roberts
et al. 2020).

On the Verge of the Quantum Ecology

In this final section, we provide some broader reflections in two complementary directions: one tackles the governance connecting actors, factors,
and discourses of the emerging quantum technologies; the other one offers
a more speculative conceptualization of the impact of such technologies at
psycho-socio-cultural-ethical levels, thus opening the way to the discussion
of the quantum ecology in chapter 5. On this, we follow up on Harro van
Lente (2000, 43), according to whom "technological futures are forceful,"
meaning that discourses and expectations on tech development contribute
to realize those same scenarios they depict.

At the core of our discussion is how we envision the synthetic operating
system of quantum technologies to operate across levels, sectors, fields, and
agencies. On the one hand, "synthesis" entails—in its meaning of "negotiation" and "convergence"—a shift from things and stasis toward processes
and equilibrium, which brings with it the need to rethink the sociotechnical paradigm as a self-organizing (geopolitical) whole, embedding, at once,
rules, roles, and mechanisms of action. On the other hand, "synthesis"—in
its meaning of "artificial"—suggests that quantum technologies will likely
create novel forms of apperception and conceptual insights. These forms
and insights will have a deep impact, legitimizing the growth and consolidation of tech-based powers, as well as debunking unfit ones, redesigning
the realm of the collective, and reconfiguring what it means to be human.

What Kind of Quantum Governance?

The important thing, from a pragmatic perspective, is to highlight trends
that can help understand how things might evolve, notwithstanding external, unpredictable shocks.

First off, the US, the EU, and China will by no means be the only players
to enter the quantum ecology. In fact, other regimes—either public or private, quantum or not quantum-specific—are already there and will appear,
similar to what happened with the internet, where language-specific and
digital-specific actors had to find a new balance (e.g., traditional publishers

and media broadcasters, alongside companies like Google, Meta, Alibaba, and Tencent). This means not only that these actors are already weaving a thick multilayered system of power, but that such a system will build on top of those characterizing the previous ecologies, as well as be disrupted by new players and technolegal arrangements, creating tensions that will remold ongoing geopolitical and market relations.

In this respect, it is possible to foresee that quantum regimes' agendas will likely enter into competition, and they will clash and/or synergically converge depending on contingencies and opportunities. For one thing, the characterization of how data regimes came into being and operate—which we explicated in chapter 3—tells us that the situation is ever evolving, and nobody can claim to safely have the upper hand. Quite the contrary, the interpenetration of digital and quantum ecologies will shake certainties to their roots. For sure, however, the quantum ecology will also develop its own antidote: *dispositifs* always contain their own resistance.

Within such a fluid scenario, we believe it is mandatory to call for an international governance on QITs able to establish shared principles and rules concerning their development and use. While regulation on QITs risks being entrapped into the "Collingridge dilemma" (i.e., to arrive either too early or too late), a "precautionary approach" to quantum techs is advisable (Taylor 2020), as well as the need to maintain a global outlook from the outset, due to the systemic and disruptive impact of these technologies. A first attempt, in this direction, has come from the World Economic Forum's Quantum Computing Governance Principles,[23] although these principles remain at a high level of abstraction. A more comprehensive and robust contribution is the one by Elija Perrier (2022, 2), who designed a "quantum governance stack" describing "benefits and risks of QITs as they affect stakeholder rights, interests and obligations across a hierarchy of international, national, public and private contexts." While the model provides valuable guidance on how to address multistakeholder governance-related issues in connection to QITs, it is normative and path dependent in nature, missing to explore paths that build upon the recognized uniqueness of these technologies.

An inspiration along the line of what was done with the Nuclear Non-Proliferation Treaty—although certainly not the only option—might come from Nash equilibrium theory, which ensures the establishment of an overt common ground—known to all actors—requiring to not deviate from agreed strategies, in view of an optimal outcome for all. This would *at least* represent an acknowledged benchmark against potential QIT-related deep state drifts and/or colonization from the outset of the quantum ecology. Dealing with national security-threatening technologies, shared mechanisms might be advisable concerning the use of *certain* technologies in *given* contexts and for *certain* purposes. Incidentally, decision theory and game theory are also fields where the application of quantum-like formalism has been successful (Busemeyer and Bruza 2012; Zabaleta et al. 2017) and might constitute a valuable reference.

Ideally, the new quantum governance will require a double articulation, defining shared principles—for instance, avoiding rivalry or mutual interference in the development and use of QITs—which are global by default and have then a local implementation by design, according to the different technocultural fields in which the actors operate. This governance framework will need to be binding, to the extent to which only the conjoint effort of all actors would maintain the equilibrium. For this reason, its enforcement shall be guaranteed and monitored by independent parties.

Living in a situation of multipolar fragmentation and considering the profound impact QITs will have, it is important to establish the conditions for a new balance. The leveraging force of AI and QITs—and likely their merging—cannot be left to the discretion of single unscrutinized actors. Since today, QITs need to be included in the *global* political agenda, alongside climate change, sustainability, and health threats, as a major driver to change over the next century.

Toward Competing World-Sensing?

Let us suppose that science had finally gotten if not to the bottom of reality, at least to the manipulation of its basic building blocks and to its evanescent temporality and spatiality. What would that change to people's daily perceptions, actions, and life?

In a quantum physics perceptual mode, the world is summoned like a thought. It is summoned by awareness and offers only what that awareness is ready to include at that moment. For sure, a "quantum-cracy"—the power of quantum—will not see the light before the next ten to fifteen years, and yet, the second half of the twenty-first century will likely greet capillary applications of QITs, calling for the need to envision and assess the impact these will have on societies and cultures. Hence, what might be the psycho-socio-cultural-ethical effects of QITs on the individual and the collective? Even though this might sound as a very speculative question—and we certainly do not pretend to provide any conclusive predictions in this respect—the very fact that the EU has warned about the possibility that QITs will conflict with principles of sovereignty and self-determination is worth exploring. As James Der Derian and Alexander Wendt (2020, 401) acutely point out, "If history is to be the judge, new asymmetries will result with the emergence of new 'quantum haves' and 'quantum have-nots.'" These tensions need to be brought to light and explored. What we can try to do, then, is to highlight deeper trends of change, starting from an evaluation of how QITs work and how they might affect the way people apprehend {reality} (cf. also Perrier 2021; Possati 2023).

To begin with, increasing quantum computing power will bring increased modularization, transferability, and replicability of experience, intended as enveloped sets of rules, variables, and embedded values to be performed in dislocated scenarios. Quantum computing, then, will allow moving a step further toward the technologization of the *ethos of the real*, as a synthesis of (individual) minds and (collective) bodies. We might say that, whereas until now experience has resided within and across subjects, with the advent of QITs, it will be increasingly diffused and instantaneous, involving both biological and nonbiological entities. And should people not modify their behaviors according to this shift, it might well be a quantum AI to take over and nudge them into responsible behaviors (cf. Simanowski 2019). At the same time, such emerging co-subjectivity will depend upon precise technocultural fields: for instance, the quantum internet that, say, the EU and China will develop will likely have different features, ruling logics, driving epistemologies, and applications, especially if the attempt at global governance fails. Insofar as QITs will pervasively impact many disciplines, the risk is to see the coming into being of competing self-enclosed world-sensing.

We already know that major applications of quantum computing will be in the realm of chemical, biological, and physics simulations. Hence, in a world of increased simulations, to what extent will humans be able, or even allowed, to comprehend and contest such renderings? Very little, seeing how blindly today people already trust and rely on data. Possibly, this might lead to a further weakening of what makes "the human" (i.e., imagination, memory, embodiment). Experience will not be the best teacher anymore: people will be led to think that simulation is (cf.. Baudrillard 1994). So, how will the consciousness of collectives increasingly relying on simulations be affected? It might well turn into a deferred consciousness, in the spirit of Plato's myth of ideas, one that is based on different layers of apperception, *including* its impossibility.

And yet, such simulations might, in the longer run, become more accurate and more vivid than those that human senses already can provide, thus drawing humans into a hyperrealistic dimension not accessible via the senses alone, but only—possibly—comparable to deeper/altered forms of consciousness. Such scenario—de facto a metamorphosed apperception of {reality} that is neither less, nor more, but gets transfixed and repeatedly reworked via a techno–human–environment symbiosis—will require sociopsychological adjustments both individually and collectively, insofar as it will be based on different encoding procedures and processing of signification.

This will not mean the end of the real; quite the contrary, there might be a resurgent attachment to phenomenological reality, but the risk is that the wrapping and molding of {reality} via QITs will further dissolve the common ground—the referent—on which to agree and act. In fact, assuming the scenario gets more polarized, it could prompt the emergence of irreconcilable world-sensing, posited on different apperception and evaluation of what is to be considered real. Different technocultural fields will create and legitimate their own referents (be they phenomenological or beyond), likely exacerbating power asymmetries both within and across these fields. For instance, secured communication via quantum entanglement might lead to increasing "opacity" of quantum actors, bestowed with an unprecedented asymmetric communication power: as Possati (2023, 12) writes, "the opacity of quantum communication is not only epistemic, like that of classical AI, but ontological." In this respect, a multipolarized sociotechnical paradigm, hosting a variety of technocultural fields, each one with its own space-time and individual–collective conditions of existence, might lead to an extremely *agonistic* scenario. To be sure, the presence of multiple world-sensing is not negative per se (in fact, it is a healthy condition); problems arise when such a scenario can no longer be rooted to a common onto-epistemological ground following a consistent—not necessarily cohesive—heuristics (the reverse—equally dangerous—is a global world-sensing without sufficient heterogeneity).

QITs materialize the extent to which relations, practices, trust, and meaning-making are emergent and shared constituents of {reality}. People will need to increasingly be able to master the *certainty of the uncertain*, that is, to deal with the {real} not only gnoseo-logically (this belongs, indeed, to the times of the privatization of the mind) or phenomenologically (collective sense-making) but also onto-epistemologically (adapting to and accommodating the tech world-sensing). On the one hand, life will become extremely "suspended" from a human point of view and yet also extremely "straightforward" from a pragmatic point of view; on the other hand, QITs will help to have access to new potentialities for the experience, thus enabling new ways of acting within {reality} beyond human faculties.

In a world of increased material and spiritual impermanence, philosophies that teach the balance of opposites, moderation, and learning to "let things go" as well as transgenerational pacts and endeavors aimed at collective long-term sustainable achievements, will adapt more easily than assertive philosophical thinking that places change within the single human mind and body. The new quantum condition will require people to make sense of the fact that the human factor is just one component of the whole system, certainly not the master. The shape that such awareness and the new sociotechnical paradigm will take is still to be seen. In the last chapter, we will dig into that, suggesting a few paths for an open (in the Popperian sense of the term) quantum ecology.

5 Ecology: A Possible Quantum-Based Paradigm

Decoherence is always an approximate notion. Complete decoherence is impossible. Indeed, if we wait a very long time, decoherence will always be reversed, as the information needed to define superpositions seeps back into the system from the environment. . . . Decoherence is a statistical process, similar to the random motion of atoms that leads to increases of entropy, bringing systems to equilibrium. These processes appear to be irreversible. But they are actually reversible. . . . The second law of thermodynamics, according to which entropy probably increases, can hold only for times much shorter than the Poincaré recurrence time. If we wait long enough, we will see entropy go down as often as it goes up.
—Lee Smolin (2019, 101)

The truth of quantum reality is an ambiguous truth. It calls upon us to live with the possibility of other possibilities. But such ambiguity can be creative. Self-organizing quantum systems thrive on ambiguity. . . . If a self-organizing system becomes too static, it runs down; if it becomes too chaotic it breaks apart. It needs a creative balance of order and chaos to thrive. This is ambiguity.
—Danah Zohar and Ian Marshall (1994, 159)

Pillars and Cement

The idea we develop here is that the quantum ecology is characterized by three interdependent pillars—uncertainty, entanglement, and discreteness, as foundational aspects of quantum physics—that come to define a self-organizing world-sensing cemented together or, better, enacted through embodiment, which, in turn, operationalizes a synthetic operating system deeply disruptive of today's scenario. Speaking of "observership," Hertog (2023, 142) writes that this act is "an agency operating at a deeper level, an indispensable part of the continual process through which physical reality comes about." In/on this continual process, QITs will exert a game-changing role.

Uncertainty

Humanity lingers on uncertainty; humanity *lives in* uncertainty. This uncertainty is not (only) epistemological or historical (the uncertainty about the future or the past), it is constitutive of life, ontological. It has, indeed, the character of epiphany. This uncertainty, this impossibility of full determinability and accountability at every moment (one could also say the impossibility to have a complete self-validating account) is what throws humanity—especially a secularized humanity—into disbelief.

Regarding increasing decision automation, Louise Amoore (2019, 149) writes that "the meaning of doubt should be reconsidered. . . . Doubt in this alternative register is felt, lived, and sensed as embodied actuality." Doubt then becomes a grounding principle that permeates experience. And yet, such a principle not only stands for the absence of certainty but also is a structuring logic of physical reality. On the one hand, this means that such uncertainty can be turned into a method, an epistemological disposition. Feynman (1994, 239) said as much when declaring, "I can live with doubt and uncertainty and not knowing. I think it is much more interesting to live not knowing than to have answers. . . . In order to make progress, one must leave the door to the unknown ajar." To know that one does (not) know is (Socratic) self-awareness: it is what entices quest, what maps the realm of knowledge *and* ignorance and provokes imagination starting from a condition of deficiency, which is key to any interrogation. On the other hand, one should never forget—and Feynman was well aware of that—the ontological foundation of such uncertainty. Humans cannot but be uncertain in the sense of open-ended. To be uncertain is to accept the implicated multiplicity of the present, and to seek knowledge is to accept the intrinsic limitation of all knowledge, which is, indeed, what makes the quest possible in the first place.

The real, then, is not a thing or a state; it is a subtraction; it is one possible onto-epistemological "revelation" (Heidegger 1977), based upon any given *dispositif(s)* and operating system(s). Knowledge and ignorance are, yes, two sides of the same coin—life and death—but information processing—at the core of any self-organizing system—is the membrane sealing their consubstantiality *and* incommensurability. The point is: such information processing is adaptive, open-ended, and intrinsically unmappable;

it cannot be imposed or hetero-directed. This is why Taylor (2020) warns that, in order to produce a politics of care, to know is not enough; it is necessary to problematize knowing as a practice linking observer and observed. This means not only to acknowledge the embodiment of all knowledge, but to identify its blind spots through a meta-reflexive process. It is here, at the juncture of the meta-reflexivity on embodiment, that a hiatus is created producing a four-square matrix, coupling, on the one hand, (a) "knowing of knowing" and (b) "knowing of not knowing" while, on the other hand, (c) "not knowing of knowing" and (d) "not knowing of not knowing." The first coupling implies and demands responsibility; the latter entails a sort of naive and fatalistic approach—yet both couplings have practical effects on the individual and the collective.[1] Foucault's (1988, 19) work on the "technologies of the self" is useful here. He drew upon the ancient Greeks' philosophic tradition, distinguishing between a "care of the self" and a "knowledge of the self":

> The precept "to be concerned with oneself" was, for the Greeks, one of the main rules for social and personal conduct. . . . [Today] When one is asked: "What is the most important moral principle in ancient philosophy?" The immediate answer is not "Take care of oneself" but the Delphic principle, "Know yourself." Perhaps our philosophical tradition has overemphasized the latter and forgotten the former.

An ecology *of* care is one in which subjects—individuals *and* collectives— freely orient their own quest within and through the world, rather than being taken care *for*. And they do so consciously. Freedom requires responsibility precisely because it implies a degree of uncertainty: while any attempt to zero either freedom or uncertainty is doomed to fail, lack of self-reflection on the affects and effects of knowledge and ignorance in the world can have disastrous consequences.

Entanglement

The condition described above is universal in that it denotes physical reality for what it is, regardless of who/what the knowing agency is. It is, then, global not merely "geographically," but systemically. It is nonlocal, pertaining at every moment to the whole, all at once. Life erupts and it is entangled: it connects by implication, it subsumes, reverberates through, and reflects the whole of which it is an instantiation. As Barad (2007, x) notes, "To be entangled is not simply to be intertwined with another, as in the joining of separate entities, but to lack an independent, self-contained existence.

Existence is not an individual affair," and she continues pointing out that "the entanglements we are a part of reconfigure our beings, our psyches, our imaginations, our institutions, our societies." This is a crucial statement that contains a critique of the objectivist idea of reality.

In fact, the notion of "intra-action" introduced by Barad (in contrast to the usual "interaction," which presumes the prior existence of independent entities) is key in materializing the idea that life is an entangled (diffracted) affair as the ever-reconfigured and actualized codependency of space—time—matter: "it is through specific agential intra-actions that the boundaries and properties of the 'components' of phenomena become determinate and that particular embodied concepts become meaningful. . . . Indeed, it is through such practices that the differential boundaries between 'humans' and 'nonhumans,' 'culture' and 'nature,' the 'social' and the 'scientific' are constituted" (Barad 2007, 101–103). Barad, in other words, considers an entanglement by default of reality, which is constantly reworked *from within* as an endless "intra-action" that literally creates reality-in-the-making, intended as a whole emergent phenomenon. Barad (2007, 103) further makes this point clear, writing that "the world is an ongoing open process of mattering through which mattering itself acquires meaning and form in the realization of different agential possibilities."

Entanglements, then, are not really "what happens" but, more radically, instantiations of all that "is." For the very fact of living, all organisms are individual entities (disentangled), yet this unicity is emergent: at an implicated level, organisms are all connected. This does not mean that everything is continuous—beings are not all one and the same—but that they are codependent. Entanglement, in other words, can be regarded as a form of codependency of all that exists, which does not negate the possibility of being as individual instantiations, but sees this individuation—what Bohm ([1980]; 2002) calls "fragmentation"—as an explicated mechanism from an underlying order of mutual implication. Of course, this codependency can be loosened, straightened, torn apart in the same way as quantum fields can be more or less entangled: this is actually what humans do all the time in this epoch of incessant destruction of the environment.

Thus, reality is an actualized emergence that gets oriented by its potentials. In his work *Untying the Gordian Knot*, Eastman (2020, 23) shows well the extent to which "the 'real' is constituted by both the actual and the possible": while (quantum) reality is entangled by default, the detection of cause–effect processes along a local–global scale is an epistemological construct, a proxy dependent upon humans' phenomenological grasping of reality. Differently said, classical physics is a limit case of quantum physics: "What is more fundamental, the classical or the quantum realm?" asks Jesper Grimstrup (2021, chap. 19): "Clearly, the answer must be the quantum. The physical reality is primarily quantum mechanical and the classical world, which we experience, is an emergent effect." In a similar vein, Kauffman and Roli (2022) write that *reality consists in ontologically real possibles, res potentia, and ontologically real actuals, res extensa, linked by measurement.* Shifting away from "measurement," we might suggest, more radically, that embodiment is, *at all times*, what is being actualized.

From here, Kauffman and Roli arrive to argue that the "brain-mind is partly quantum." Even though their understanding of intelligence is still very much anthropological, the key adverb "partly" is relevant: the idea that cognitive and perceptual processes might be *partly* quantum-like opens to the possibility of a hybrid interpretation of these processes (from which the authors' "trans-Turing machine" derives), getting rid of micro–macro and classic–quantum dichotomies. Life itself is a codependency of thought and action: thought is already action, and action is always already a cognitively embodied practice: an organic-*dispositif* enframing.

Discreteness

Physical reality is discontinuous, discrete, and this is exactly what makes it ontologically generative. "Nature has a fixed number," Grimstrup (2021, chap. 4) indicates, "which tells us that there is a certain amount of interaction that we can never avoid. One of the first consequences of this unavoidable smallest interaction is a radical shift from . . . numbers to the process of obtaining these numbers." Indeed, such discontinuity—identified by Planck's constant—determines a basic condition, a constitutive *écart* of reality, which inevitably presupposes a separation, a fracture into the whole by which and through which reality is what it is. Gustave Courbet was right: *L'origine du monde* is a hole; at the bottom of life is a nonorigin. This discontinuity is infinitely reworked in kaleidoscopic ways whereby the particularities making up totality cut through this latter and actualize it. *Tout se tient*: past and future, here and there, "I" and "Other." Everything *is* generative (by) differentiation: nature does not exist as a given; it cannot stay still. This *écart* is not only a "gap" (cf. Zellini 2022) but a *limen*—a (in)tensed vibration—that cuts through the continuous to create the discrete as endless possibility: the *écart* "forces" the discrete into being. Nature has found the cheapest and smartest way to "allow" (multifarious) existence: this way is neither to be—which would presuppose all sorts of specifications and attributes—nor nothingness—which only exists analytically—but an in-betweenness.

Quantum field theory teaches us that physical reality is always somewhat excited even at a minimum level of energy. The physics of the *écart* presupposes a restless dimension in which, at once, time and space, energy and matter, coalesce, and also where these originate from: "These ordinary notions," Bohm (2002, 21) contends, "in fact appear in what is called the explicate or unfolded order, which is a special and distinguished form contained within the general totality of all the implicate orders." The *écart* then has no positivist ontology as such; it is an encoding emergence that renders information pertinent as a differential process, in the spirit of holographic physics: "holography ingrains a fundamental element of emergence into the very roots of physics" (Hertog 2023, 175). This is a fundamentally Darwinist idea of physics (and its laws), which considers evolution as dependent upon a degree of casualty. When one speaks of intention, the tendency is to anthropomorphize this concept; de facto, however, intention is just a tension between the possible and the impossible: there are no projects in nature in the traditional sense of the term: what is there is a fundamental open-ended orientation of possibility(ies).

What one perceives as physical reality is holographic (cf. Bortoft 1985; Velmans 2000) to the extent it refracts the entirety of which it is constituted and constitutive; better, it is a holographic *practice*, meaning that experience always contains the whole within itself and it is only the bodily individuality that pins down the focus of such grasping to a single point-of-being, which, however, is always already inscribed into the whole. Being reaches out to the Other for the very fact of being; in fact, beings *are* because they reach out; they centripetally recall into their-self all that surrounds. This is not just some philosophical meandering; it is physical reality as described by quantum mechanics: to be is to act (and to know). Uncertainty, entanglement, and discreteness are the founding onto-epistemological principles of the quantum ecology.

*   *   *

As physicist Smolin (2019) notes in the excerpt, the physical reality oscillates between decoherence and coherence (over a very long time frame!)

and, according to Danah Zohar and Ian Marshall in the second excerpt, so do individuals and society at large. The equilibrium that coherence implies is not some sort of congruent harmony. In physics, a system in equilibrium simply means that the behavior of its parts is not arbitrary and can be calculated and measured. Socially speaking, a coherent order is one in which the actors at play are caught in mutually implicated power relations, so that the whole order is sustainable/sustained for and by all actors. This understanding of course does not exclude forms of power asymmetries: elites and subalterns are and will always be there, always enacting ongoing struggles for mutual determination. The language and digital ecologies predicated and established their respective coherent world-sensing; the quantum ecology will establish its own. From what we have discussed above, the quantum ecology is fundamentally *communitarian*, although not in the sense this term is usually understood. People tend to think of community as a "positive" concept, a gathering together, based on common goals (or even properties), but in fact community—as Roberto Esposito (2022) describes nicely—is a "negative" concept: it derives for *cum + munus* ("duty together"), denoting a bond among people based on *necessity*. In other words, the communitarian bond rests on a gap, a fundamental condition of deficiency. This must also represent the starting point when discussing the quantum ecology: humanity can be self-sufficient only as a whole; trade-offs must be rebalanced reiteratively by remaining open to change and adaptation. Mobilization in this direction has already been called for: "[The] growth of research on quantum technologies calls for a societal debate to explore and assess the impacts that quantum technologies will have on science, industry, people and society" (Vermaas 2017, 241). A collective effort to envision together such kind of quantum ecology is needed.

Quantum Ecology

It is fairly safe to envision a *crisis*—in its etymological sense of "decisional turning point"—in the next ten to fifteen years as the result of the misalignment of language, digital, and quantum ecologies. The quantum ecology is already affected by actors, factors, and discourses, yet its outcome—the shape of the new paradigm—is not. Brian Massumi (2017, 353), in his article "Virtual Ecology and the Question of Value," points out that "the braiding [of causality] is nothing like a mechanical part-to-part connection. It is a cooperation, across the differential between the objective and the subjective,

and the actual and the virtual, that brings the occasion to life." This resonates with what we discussed in chapter 2 about the grasping of information
as an embodied process. But how to proceed from here, moving beyond
the mere proliferation of and call for new sensibilities? How to bring about
profound and radical change? "The solution," Floridi (2020, 120) says, "is
a human project that is not only a meta-project caring for individual projects, but also a collective project." To get there—we believe—a quantum-
based transformation of perception, method, and action must seep into
global consciousness.

*   *   *

While drafting this last chapter, we have wondered about the signs and
symptoms of the emerging quantum ecology. We have looked for an event,
not necessarily big, but paradigmatic, that could help us guide the reader,
in the same vein as the launch of the first PC by Apple in 1984 or the fall
of the Berlin Wall in 1989 signaled the emergence of the digital ecology. In
this case, the quest was rather simple but only apparently banal: the Covid-
19 pandemic has really represented a major turning point.

First off, the pandemic has demonstrated that today's global society is
mainly so, if not exclusively so, on a material level. The merciless logic of
productivity knows no pity, nor remembrance, and in fact, it only produces a self-isolating connectedness. The pandemic has reminded people
that collective life comes with rights and mutual duties. The fundamental
sin of materialistic connectedness is to have eroded a vision of and for
the collective. To be entangled is not only to be connected but also sharing the same fate. Like it or not, humanity *is* Covid-19: it is a distributed
reverbero-manifestation, all at once, everywhere, of humans' actions in
the world.

Next, the pandemic has left humanity bare-naked in front of uncertainty: people's pretension to get rid of uncertainty (through calculability)
has been shattered in one blow. As Taylor notes (2020, 3), the competing scientific truths about the pandemic "warn that 'the data' on Covid-
19 does not exist. . . . Instead, the pandemic has revealed the scientific
method in all its socio-technical, Latourian chaos," by which she points
to the ever-partiality of any given ecological paradigm. In this context,
the many competing truths about the pandemic are symptomatic of the
fact that the datafield, per se, cannot ever be self-validating or deliver

ungrounded "right" answers. Indeed, the pandemic-induced uncertainty that humanity was forced to face is not only of an epistemological nature—doubting data and scientific methods that best account for /reality/—but ontological. Truths were "competing" because they were decontextualized and instrumentalized: what was missing was principles to drive the reading of data *through* the crisis. The benefit of doubt for the sake of the collective—buried under frantic acceleration and rivalries—went totally amiss.

In the next section, we will characterize the features of the quantum ecology in terms of space and time, as well as (diffused) subjectivity. In the subsequent section, we will then dig into a number of fields, advancing examples that inflect and particularize the quantum ecology, being aware, however, that only synergically tackling socio-economic-environmental issues can really make a difference. The quantum ecology is a meta-ecology.

Main Features of the Quantum Ecology

> The fact of our belonging to this moment at which a change of epoch, if there is one, is being accomplished also takes hold of the certain knowledge that would want to determine it, making both certainty and uncertainty inappropriate. (Maurice Blanchot 1993, 262)

Blanchot's words ghostly recall quantum physics and quantum ecology as discussed throughout this book, notably with regard to certainty and uncertainty, change of epoch as a kairological moment, and determination as an embodied practice, that is, in essence, *what* we can know, *how* we can know it, as well as the very essence of this *we*. Most important, from within a certain ecology, these issues cannot be recomposed into a complete and self-validating scheme. Blanchot states clearly that it is the radicalism intrinsic to any epochal change that, by undermining the ontological and epistemological foundations of what preceded, provokes a blurring of un/certainties, a sense of disorientation and inappropriateness, which can only be dispelled, or at least relativized, from an outer/different perspective. As Stiegler (2017, 136) contends, a "technological shock is epochal in as much as it makes an epoch, that is, it is a suspension, an interruption, a disruption, and as such stupefaction." In other words, the advent of a new set of technologies has an epiphanic impact on reality, which reverberates throughout (re)defining the horizon of possibilities. Stiegler continues arguing that the "epochal technological shock is stupefying in that it disrupts the organolog-ical arrangements established by a prior and meta-stabilized stage." In fact, it is the concept of certainty itself that gets contested: while an epoch always strives to make sense of itself, it can never fully achieve such a goal. To make sense of the change of epoch, it is necessary to move beyond its own onto-epistemological horizon of existence, and to do that, new lenses are needed. Hence, what *kind* of embodiment will the quantum ecology enact? How will QITs contribute to that?

To answer these questions, it is important to think through the quantum ecology, both cognitively and programmatically, by reformatting what is to be/act/know in the world. Fortunately, signs and traces that go in this direction can already be spotted, although they need to be synthesized into a coherent picture.

Reconsidering Space: Nonlocal Differential Space

To speak of space in relation to the quantum ecology, an experiment of quantum mechanics is worth discussing: this is the "bomb-tester," a thought experiment elaborated in 1993 by physicists Avshalom Elitzur and Lev Vaidman (1993) and then performed in the laboratory in later years. Basically, the experiment builds upon the wave–particle duality to perform an interaction-free interrogation. Physicist Sabine Hossenfelder describes it as follows on her webpage[2]:

> Suppose you have a bomb that can be triggered by a single quantum of light. The bomb could either be live or a dud, you don't know. If it's a dud, then the photon doesn't do anything to it, if it's live, boom. Can you find out whether the bomb is live without blowing it up? Seems impossible. But quantum mechanics makes it possible. Here's what you do. You take a source that can produce single photons. Then you send those photons through a beam splitter. The beam split-ter creates a superposition, so, a sum of the two possible paths that the photon could go. Along each possible path there's a mirror, so that the paths meet again. And where they meet there's another beam splitter. If nothing else happens, that second beam splitter will just reverse the effect of the first, so the photon continues in the same direction as before. The reason is that the two paths of the photon interfere like sound waves interfere. In the one direction they interfere destructively, so they cancel out each other. In the other direction they add together to 100 percent. We place a detector where we expect the photon to go, and call that detector A. And we put another detector where the destructive interference is, and call that detector B. In this setup, no photon ever goes into detector B.

But now, we place the bomb into one of those paths. What happens? If the bomb's a dud, that's easy. In this case nothing happens. The photon splits, takes both paths, recombines, and goes into detector A, as previously. What happens if the bomb's live? If the bomb's live, it acts like a detector. So there's a 50 percent chance that it goes boom because you detected the photon in the lower path. If the bomb is live but doesn't go boom, you know the photon's in the upper path. And now there's nothing coming from the lower path to interfere with. So then the second beam splitter has nothing to recombine and the same thing happens there as at the first beam splitter, the photon goes both paths with equal probability. It is then detected either at A or B. The probability for this is 25% each because it's half of the half of cases when the photon took the upper path.

In summary, if the bomb's live, it blows up 50% of the time, 25% of the time the photon goes into detector A, 25% of the time it goes into detector B. If the photon is detected at A, you don't know if the bomb's live or a dud because that's the same result. But if the photon goes to detector B, that can only happen if the bomb is live AND it didn't explode.

*   *   *

What does this mean? In simple words: quantum mechanics allows one to know "something," even though this something did not happen/was not observed (i.e., the path the photon did not take). More deeply, through a sort of insightful abduction expressible only through math's formalism (natural language would imply its expression and immediate erasure), the experiment attests to the nonlocality of quantum physics, bearing valuable insights for our spatial conception of the quantum ecology, as a dimension that must be regarded as simultaneous and that demands to be approached multiperspectively, seeking a differential balance able to accommodate and adapt to local uncertainty.

From an ecological perspective, valuable paths for reconsidering space, people's relation *to* it, their presence *into* it, and, more radically, their coming to being *together with* it, come from the Japanese and Chinese cultural traditions. Traditionally, both these cultures have relied on a qualitative conception of space that not only departs substantially from the Western idea of neutral immutable space but also is connected with a qualitative conception of time and considers the two conjointly.

For the Japanese, space is a continuous flow, alive with interactions and ruled by a precise sense of timing and pacing. The name for that is *ma*. To the Japanese, *ma* connotes the complex network of relationships between people and objects. A French expert on Japan, Michel Random (1985, 149–150), notes,

In Japan, everything depends on ma: the martial arts as well as architecture, music or plain art of living. Aesthetics, proportions, garden design, all belong to networks of meanings which are related to each other through ma. Even business people in Japan obey the laws of ma when they approach each other; the idea is to sense how your partner judges things. Ma will then dictate the hierarchy of choices, the priorities of the investments, the right time and the proper pace in the organization of the enterprise, and shape the exact perception of people and situations. In a word, ma is perceived behind everything as an undefinable musical chord, a sense of the precise interval eliciting the fullest and finest resonance.

Simply put, *ma* regulates spatial harmony. Most important, it does so via an immanent standpoint that foregrounds relations over relata. *Ma* is a qualitative differential space in which nodes and edges are not flattened onto a mere rhizomatic ideal; rather, it is distances—the "white holes" within space—that define space's structuring logic and connote it as a harmonic totality. The Japanese have a keen awareness of the interval. This is reflected in Japanese design, for instance, flower arrangement, gardening, or hierarchical interactions. It is the whole that presupposes its constituents and their relations, or better, spatial harmony is substantially different from all that summons its materialization.

As people develop increasingly embodied technological interfaces to negotiate between their point-of-being and the world, it is necessary to promote "differential spatial thinking" (as per the distinction we made previously between three spaces), a safeguard to keep them aware of how they occupy reality, just as Jean-Luc Godard would insert clumsy breaks into the continuity of his films to remind people they were watching a film and not get caught into the illusion of the real. This leads to conceive of the human relation with technology according to a sort of psychotechnological *ma*, a world of Janus-faced (technohuman) intervals in constant mutual dependence. Based on psychotechnological *ma*, a sense of proportions within the dictatorship of the *datum*—considered as the basic ratio for a /reality/ that has lost sense and meaningfulness—is restored. Technology is de-fetishized as a tool and inscribed into a qualitative space for which it does not function anymore as a yardstick of command and control but as an organic component that demands relationality in the form of a biotech consubstantiality. Since *ma* is the interval between items, objects, people, and entities, it makes sense that the interval between user and system, that is, any interface, or other manner of interactivity provided by the technology, is a kind of *ma* too. And it is appropriate to think of such *ma*-dependent technological interfacing as embodied in nature in the first place. Just as an example, through a project named "Grow Your Own Cloud,"[3] it is now possible to take digital data, translate them from binary code into DNA, and then, through synthetic biology, store it in plants that can still capture CO_2. While the drawbacks at scale of such an approach must also be explored, this can be a way to restore a balanced relation between technology and the environment—as a technology-through-environment configuration— where the "human" does not set out to command and control either one or the other but seeks a synthesis with them.

Similarly, Chinese traditional culture has the concept of *feng shui*, which literally means "wind" and "water." *Feng shui* identifies the way to find and keep an overall balance in the organization and occupation of space. *Feng shui* is the Chinese way to conceive and inhabit space holistically, more than geometrically, an attempt that reflects the Chinese cyclical approach to the cosmos compared with the linear, rationalistic one of the West. In the words of Alfred Hwangbo (1999, 191–192),

> Feng shui can be defined as a mélange of art and science which governs design issues of architecture and planning, embracing a wide range of disciplines. . . . *Chi*, the vital cosmic current which runs the universe and also means "breath," can be scattered when it meets wind, and can be stopped when it meets water. For East Asians, the substance or rather energy *chi* was the basic constituent of all things, a Chinese alternative to the atoms of matter assumed by the Greeks and inherited by Western science.

While *ma* is an issue of focus, of paying attention to interval, *feng shui* is one of orientation, paying attention to positioning. In fact, it mixes point of view and point-of-being, hard and soft disciplines. Note also that the Chinese idea of *chi* resonates with that of wave in modern physics, while the Western idea of the atom, as seen, is tied more to a particle-like conception of matter. Hwangbo (1999, 194–196) continues specifying that:

> The main instrument of feng shui practice is the magnetic compass. Feng shui compasses do not simply register the cardinal points like Western ones, but are equipped with a complex series of circles inscribing various kinds of information from different ordering systems . . . Practitioners of the compass school would read the compass on site, noting the relationship between natural features and the directions registered by the various compass rings. . . . The aim was to organize the built environment in harmony with nature.

Here one gets a clear description of an integrated and qualitative conception of space: integrated because it not only depends upon but also posits the mutual codependency of elements—be they physical or conceptual—as the precondition for a harmonic organization and inhabiting of space; qualitative because such integration is not merely based on the metrics of space but foregrounds the energetic and mattering consubstantiality of elements. These two assumptions are then put into practice through design and architecture, as manifestations of the harmony that *feng shui* seeks to achieve, intended, above all, as a value-laden quest.

Today, also in the West, design has evolved from an essentially reactive to a more proactive stage, implying a design that does not simply map space but attunes to it. Accordingly, technology, as a built-in component of people's relations in the world, is turned into an object of design, rather than being at the source of design. From here, design also evolves into exploring and creating patterns of interaction and interfacing beyond the production of objects, an issue at the core of what is discussed above about the sustainability of biotechnology. Similarly, it is not surprising that from various parts, tech experts have already warned against the risk of "naturalizing" (i.e., humanizing) AI. One can neither consider the human–technology–environment relation as the result of a triad nor blur all differences. Harmony is the balance emerging from/through differencing. A conception of space that recognizes differentiality, while seeking an immanent orchestration of the parts defined by it (and repeatedly redefined by it), leads to the coming into being of a balanced ecological dimension. (F)actors are mutually interdependent at all times, *even* in the absence of direct local connections: as a situated practice, networking alone does not by itself make up and account for the sustainability of the system—it demands to metamorphose into a systemic procedural conception, enacting an evolving entangled ontology.

Hence: how to conceive and design, for instance, symbiotic autonomous systems according to *ma* and *feng shui*? *Feng shui* brings up the question as to how a personal digital twin interacting with a collectively intelligent environment could best position the physical person to take action; *ma* requires such positioning to be codependent to that of other organic and inorganic entities. The source of inspiration for design can no longer be limited to traditional notions of beauty and efficiency but must include the recovery of humanity's most ancient wisdom regarding living in a kind of space that is alive—simplicity, lightness, and flexibility will be key. Indeed, it is sensible to think that QITs will produce forms of virtual spatiality that sew distant nonlocal features together, opening the door to forms of ultra-experience where the point-of-being is spread, diffused, while the point of views are multiplied and/or coalesced. In this scenario, to follow *ma* and *feng shui* as principles will help design pleasurable, rather than distressful experiences, functioning as design compasses of the emergent synthetic spatial dimension.

Reconsidering Time: Trans-Time

As for space, also when it comes to time and the quantum ecology, it is worth discussing a famous experiment in quantum physics—the "delayed-choice quantum eraser" experiment—not much for its (wrongly supposed) wizardry but because it helps bring the temporal contingency and immanence of any knowing-being endeavor to the fore.

First conceived as a thought experiment by Wheeler in 1978 (cf. Marlow 1978), the experiment has been empirically conducted, in slightly different setups and by various research groups, throughout the 1980s, 1990s, and 2000s. The delayed-choice quantum eraser experiment is an evolution of the double-slit experiment, and one possible setup is the following: a beam of photons passes through a double-slit barrier. After the barrier, a crystal creates a pair of entangled photons—*signal* and *idler*—from each photon passing through the slits. What the crystal does is to convert the photons from either slit into two identical, orthogonally polarized entangled photons. Of the entangled pairs created, *signal* photons are sent to hit a detector (D0), where they show particle-like behavior because through entanglement the information about which path (slit) they took has been observed. *Idler* photons, instead, are directed to two beam splitters (BS1 and BS2). Beam splitters are devices that transmit light of a certain polarization and reflect light of orthogonal polarization. So, one half of the *idler* photons passes through and the other half is reflected: those that pass through BS1 are reflected by a mirror (M1) and then detected by another detector (D1); those that are reflected by BS1 are detected directly by detector 2 (D2); the same goes for *idler* photons encountering BS2: they can either pass through, thus hitting a mirror M2 and then be detected by a detector (D3), or be reflected and detected by detector 4 (D4). In doing so, a condition is created where it is no longer possible to know which path through the slit each *idler* photon took (the erasure part of the experiment). When the *signal* photons, entangled with *idler* photons detected by either D1 or D3, are observed, interference patterns on D0 appear as if the information on which path the *signal* photons took had been erased. The delayed part of the experiment has to do with the fact that the erasure can also occur a long time after the *signal* photons hit the screen, by simply making their path longer.[4]

To be sure, the experiment does not entail reworking the past from a future standpoint; rather, this is another confirmation of the complementarity principle, whereby the whole experimental setup determines the possible experimental predictions. As Barad (2007, 272) notes, "It's not that the experimenter changes a past that had already been present or that atoms fall in line with a new future simply by erasing information"; at the same time, however, she extrapolates and offers an insightful reading: "The point is that the past was never simply there to begin with and the future is not simply what will unfold." What Barad points to is an interesting implication of the experiment from an onto-epistemological perspective: notably, the experiment exposes the (space-time) conditions of its own interpretative framework. This, according to Barad, red flags the idea of objective givenness of time in favor of an understanding of time as an ongoing emergence that "happens" at the very moment of interaction/measurement: to know necessarily means to embed a certain point of view as well as point-of-being; it implies to take a stance—a certain embodied enframing—and to be(come) part of that being known within a shared space-time frame.

*   *   *

This brings us back to the codependency linking humans and technology discussed in chapter 1, now reframed into a broader ecological understanding. Following upon Heidegger and Leroi-Gourhan, Stiegler (1993, 2009) has been among the most original thinkers to elaborate a techno-dependent ontology of being, "oscillating back and forth between transcendental critique (metaphysics, phenomenology, deconstruction) and empirical history (evolutionary biology, paleontology, techno-science)" (Bradley 2011, 127). According to Stiegler, being is never present-to-itself in that it can only be defined—or, better, *be-come*—in relation to its past (what no longer is) and its future (what is not yet). Being has no essence per se: it is constantly retro-jected and pro-jected (cf. Wambacq and Buseyne 2012). Most important, according to Stiegler, it is *through* technology that the human being can grapple its temporality as extension toward the past and the future. Without technology—without embodiment—there is no time (or space).

This is why Stiegler (1993, 254) claims that "a tool is, before anything else, memory: if this were not the case, it could never function as a reference of significance": for Stiegler, being and doing are mutually imbricated, with the latter prosthetically defining the former. While correctly positing a codependency between technogenesis and sociogenesis, the problem with Stiegler's conceptualization of the techno-temporality of the human being is that it remains anchored to a (macro)linear understanding of time—in fact, a myth of origins—and to an anthropomorphic agency. On the one hand, the destructuring of time proposed by Stiegler, however valuable, remains metaphysical; it is just a first step that is not developed in its full implication, that is, an ever-temporality that can be grasped only trans-temporally by cutting through it. There is no origin, either human or technological. On the other hand, the sociogenesis that Stiegler depicts does not move past the *anthropos*, failing to extend the discourse to reality as such, which implies to acknowledge the technological nature of any embodiment, regardless of the agent.

Especially concerning the first point, Stiegler remains Derridean throughout. As a matter of fact, Derrida and Stiegler (2002, 12) do acknowledge together that time is always a trace, a bio-tech-dependent projection:

> There is never an absolutely real time. What we call real time, and it is easy to understand how it can be opposed to deferred time in everyday language, is in fact never pure. What we call real time is simply an extremely reduced "différance," but there is no purely real time because temporalization itself is structured by a play . . . of traces.

However, they never turn such understanding into a positive (ecological) thinking able to move past given temporal categories of signification (cf. Plotnitsky 1994). On this, Guattari (1995, 125) is more insightful when he notes that "transversality [is] never given as 'already there,' but always to be conquered through a pragmatics of existence." The linearity of time must be complemented with a kairological conception of time as the one informing the performativity of data-driven technologies, that is, an epiphanic time with no past nor future. The apprehension of such a bond between chronological and kairological time is a *trans-time* dimension that not only is immanent to reality but also *is* embodied reality in the making. Trans-time is a time of happening that can also be grasped as passing. Events happen and they impact everywhere, all the time, as nonlocal reverberations, and yet they also "come and go," meaning that agents can intervene and act in these events.

And so, what might an experience mixing kairological and chronologi-cal time be? One example comes to mind to which any reader can relate: boredom. Too often labeled as downtime, boredom is actually an uptime to the extent it is able to restore the focalization of the subject's own being-in-the-world. In fact, boredom is the (apparently empty) anteroom of creativ-ity, the precondition for a reconfiguration of the possible; it is dwelling *in* possibility. Under this lens, boredom is everything that resists and escapes efficiency and computation *as well as* linguistic expression; it is temporal extension ad libitum—neither chronological nor kairological; neither col-lected moments, nor eruption—which demands to be lived through.

To move a step beyond a pure dialectics of opposites forces the metaphys-ics of origins into a *dispositif*-dependent reality that accommodates existence as phenomenological becoming, that is, one to which openness and no par-ticular individuation can be attached. All living beings are "living" because they are technological, that is, actively branched to their own environment, endlessly processing it and being processed. To say that "in (their use), tools disappear. Their mode is being-ready-to-hand. . . . This 'ready-to-hand' actu-ally forms the originary relation between the human and the nonhuman" (Stiegler 1993, 21) is not enough; this must be complemented by a *non-human–non*human relation that dictates an implicated order of existence.

The environment, after all, is neither nature nor culture; it is a quali-tative dimension of differentiation and orchestration. From an ecological perspective, the whole environment—in its multifarious instantiations—can be, *in potentia*, the tool of itself, thus having a collective memory that gets actualized in the very act of living (together). This idea can be better grasped through another simple example: the (ab)use of smartphones to rec-ord life events, such as concerts. A trans-temporal experience would require, by contrast, to avoid having the event filtered through smartphones, not so much for an aprioristic adversity to technology but for pushing the environ-ment to the fore by bracketing any artificial interfacing with it. Letting smart-phones off allows one to "enjoy the moment," as the saying goes, but also to finetune one's own body—which is also a technology, although we tend to disregard it—with the ethos of the event, an ethos that people can enjoy as much collectively as individually. From the vantage of their point-of-being, spectators would then let music fill and dictate the time of the happening and make it cohabit within their own personal time, having their memory impressed in a way that is impossible to parcel, that is, presubjective.

In this scenario, QITs might well become enablers of trans-temporal experiences that enact a synthetic artificial time that can be diluted, expanded, contracted, fragmented, metamorphosed at will. Especially, QITs will foreground a productive sense and experience of temporal uncertainty. A socio-technical time that can be lived and relived, or even suspended, will in turn deeply impact on people's psyche, bringing forth a redefinition of their sense of self: people will live (through) manifest information coincidences and dissociations. On a more practical level, the shrinking of time for the simulation of complex systems enabled by quantum computers will likely redefine and redesign the human–environment relation toward both an increased objectification (processed life) and decreased object–subject distance (immersive experience).

Entangled (Social) Consciousness

What makes the subject an "I" are, respectively, subjectivity (or consciousness) and self-awareness (being aware of being conscious). Recently, quantum physics has been invoked as a conceptual framework for the understanding of subjectivity and self-awareness. In his *Quantum Mind and Social Science*, Wendt (2015, 137) builds a rich account based on the premise that "life is a macroscopic instantiation of quantum coherence." By this token, consciousness is regarded as the "subjective manifestation of wave function collapse in the moment, but which is also reconstituted as a stream of such moments by the protective shielding of the organism's body" (139). Wendt has a point in considering the body as a sort of permeable membrane negotiating the in-and-out interfacing with the world.

However, without endorsing Wendt's reference to the wave function collapse, the body can be seen, more efficaciously, as an instantiation that "cuts" through the entangled reality of which it is part: "we (but not only 'we humans')," Barad (2007, 355) writes, "are always already responsible to the others with whom or which we are entangled, not through conscious intent but through the various ontological entanglements that materiality entails." According to Barad, living beings exist as diffractions of the entangled whole. From this perspective, consciousness (subjectivity) is the embodiment of the (whole, potential) entanglement of which any body is part. Then, only a portion—an instant—of embodied consciousness turns into awareness: making sense of this entanglement (which does not imply understanding it) is, in fact, thinking-as-awareness, or consciousness turned aware. As the phrase "to make sense" suggests, aware thinking is an embodied "enaction" (Varela et al. 1991), that is, focalized consciousness, consciousness with a purpose or direction. This focalized consciousness guides the interfacing of (embodied) subjectivity through (entangled) reality. On the other hand, the unconscious is the embodied entanglement of reality not (yet) aware. One thing is to know; another is to know to (not) know: it is emergent awareness to open a whole realm of possibilities for *testing* the resilience of knowledge.

Of course, a subject can be aware of something without turning it into outward action. This is just a matter of more or less rational evaluation. The point to make is deeper: awareness is *already* an effect. Insofar as a subject is the embodiment of the entanglement of which it is part, by default, the subject gets bodily affected (and in turn affects) that same whole. Aware thoughts are the peaks of the iceberg, so to speak, but all that is concealed does play a role and reverberates into the entangled (outer) world—including the bodily subject—while, in turn, being informed by it. There is no need to turn to spiritualist philosophies or panpsychic approaches, however valuable these are (cf. Bitbol 2008; Theise and Kafatos 2013). The work *Consciousness and the Social Brain* by neuroscientist Micheal Graziano (2013, 14) is helpful in this regard: "Consciousness," he notes, "encompasses the whole of personal experience at every moment, whereas awareness applies only to one part, the act of experiencing." Awareness is a process, that is, an ongoing embodied focus repeatedly attuning to the entangled whole.

According to Graziano, it is attention (as an in-tension) to make a subject aware. Attention is the narrowing down, the zooming in on certain aspects of the consciousness-world entanglement; it is, in other words, an "adaptive useful internal model" to turn consciousness into awareness. Aware thoughts, then, are constantly re-created bundlings of attention, emerging from the whole of consciousness of which the body provides the in/out membrane. It is in this respect that it makes sense to speak, by default, of a social consciousness. Life experienced as subjectivity is an instantiation (and therefore a particularization) of reality: an open-ended embodiment. This is an inherently ecological conceptualization in that it implicitly dislocates consciousness from a precise or unique location or time and makes it distributed; privatization comes only *after*, as an internalization and reworking of entanglement. As Hans Bernard Schmid (2014, 7) points out, "The 'sense of us' is plural pre-reflective self-awareness"; it is something that precedes the emergence of (purposeful) individuality. Such a vision finds a reflection in the way people apprehend and act in the world (cf. also chapter 2 on the grasping of information). On this point, Graziano refers to the concept of "affordances" as hooks that light up a subject's attention, thus enacting focused thought. On a closer look, there is no need to identify discrete "stimuli" or "things" toward which attention is directed; rather, it is the whole reality, above the Planck's threshold, that can work and be repeatedly configured as a (huge) affordance, whose potentialities are unlimited and context dependent: as James Gibson (1977), who introduced the term *affordance*, states, affordances are both "a fact of environment and a fact of behavior." All physical reality is information and can be turned into relevant experience, depending on virtually every possible embodiment disposition. It is in this regard that Gabora, Rosch, and Aerts (2008, 94–95) point out that "because it is an organismically meaningful world that is perceived and acted upon, form and function are as inseparable and co-defining as perceiving subject and perceived object, and this is the information that constitutes both perception and action." Subjectivity, then, is always already *co-subjective*: everything "out there," perceived and experienced as objective, involves a totality of possibilities: features of the reality become pertinent based on peculiar cognitive–perceptual enactments as creative insights.[5]

How might the effects of a co-subjective understanding of reality manifest? Let's take a famous case: the "Rat Park experiment" (Alexander et al. 1981). To study the effects of drugs and understand if/how they lead to addiction, in the first half of the twentieth century, a plethora of studies was conducted on rats in isolated cages, to which the option was given to drink water or a solution containing either cocaine or heroin. In the majority of cases, rats opted for drugged water, leading soon to addiction and to their own death. In the late 1970s, a team led by Canadian psychologist Bruce Alexander investigated whether the chemical composition of drugs was all there was to say about the reasons behind addiction. They tested this hypothesis by creating a "rat park" where a rat colony could eat, play, socialize, move freely, and have sexual intercourses. Then, when given the chance to choose between normal water and drugged water, the overwhelming majority opted for the former. This led Alexander and colleagues to suggest that contextual environmental factors like isolation or the exposure to an insufficiently diverse environment do also contribute to foster addiction. A similar conclusion, involving humans this time, was reached by Dutch psychiatrist Peter Cohen (2009), who elaborated the hypothesis that addiction is a form of (substitutive) bond to which people resort whenever they lack a sense of personal inclusiveness, belonging, or life meaningfulness. Drugs, in this respect, are "agents of bonding" that help individuals cope with the world, in the same way as friends and healthy and/or social activities are.

More generally, these studies show the extent to which humans are *Erfahrung*-seekers, that is, they need to be involved in lived experiences that foster an ethos of collectivity. In a similar vein, just think of education during the pandemic. Statistics around the world show a plummeting of school scores during the period of isolation: sure, a multitude of causes could be factored in—from diminished attention to health issues and psychological distress. All these, however, are symptoms; at a deeper level, it is not heretical to suggest a common root to these symptoms, notably the impossibility to share, to be part of the same environment, to foster a synergic dimension of interaction and mutual learning.

In this context, the synthetic logic of QITs might contribute to materialize the idea of entangled social consciousness by *forcing* an apprehension of {reality} as co-subjective by default, that is, redesigning experience in ways that constrain the self to its (nonlocal, trans-temporal) surroundings and bond it to immediate consequences. Responsibility will not only be manifested and expressed but also will be *felt*; everybody—or, better, every body—will *sense* that the world is everyone's backyard. Even as the point-of-view attitude is being replaced by a participatory point-of-being, at the same time, through QITs, the (literate and digital) sense of self might be increasingly sealed off from scrutiny and become opaque even to oneself. As a matter of fact, this is a trend already going on through datafication, as seen in chapters 2 and 3, and might find in QITs the enablers of a synthetic self that fully escape people's grasping. One way or another, the sense of identity will be profoundly reshaped: Possati (2023) acutely notes in this regard that we will likely witness a "disintegration of identity understood as the definition of boundaries between individuals and groups," leading to rethink the very notion of privacy (cf. Purtova 2017) and agency (e.g., the reconfiguration of gendering beyond boxes)—we claim—toward glocal and collectual forms based on paraconsistency (cf. Priest et al. 2006).

Ideas for the Quantum Ecology

In the remainder of the chapter, we provide concrete examples, cutting across various fields—education, economy, tech innovation, urbanism and design, language and arts—about principles and *modi operandi* to realize an open, communitarian quantum ecology. These are to be intended more as possible paths rather than solutions; they are methodological compasses refracting the discussion conducted earlier, de facto opening questions rather than providing answers. Moreover, the list works as a guideline and is far from exhaustive; rather, it naturally calls for future integrations, which can be more or less specific to certain technocultural fields. Our goal is, simply said, to put theory in practice.

Transdisciplinary Education

To begin this survey with education is a political choice. Leaving aside discussions about the importance of giving free access to education, the big question here is: What *kind* of education (by/for the quantum ecology)? First off, the contributions of Christopher Coenen and Armin Grunwald (2017), as well as Alexei Grinbaum (2017), are valuable in that they call for an involvement of multiple stakeholders in the debate on the development of quantum technologies. However, such debate must be grounded on a shared understanding of quantum physics, which still needs to be properly fostered, especially for nonexperts. It is evident that this shared understanding constitutes the *conditio sine qua non* for avoiding the co-optation and commodification of the quantum ecology by big data regimes.

Beyond that, the pedagogical paradigm of the quantum ecology is different from those of other ecologies. The subject of the language ecology in its alphabetic mode was a person of reading and writing, a principle that was well understood by the Jesuits at the beginning of modern times. This pedagogical method—called the Ignatian pedagogical paradigm (IPP)—promotes the abilities of students by making them active learners. The IPP pays particular attention to (1) the context and the way in which students acquire knowledge; (2) experience, understood as a process of internalization and reworking of knowledge; (3) reflection, understood as the commitment to conceptualize and make sense of what was learned; (4) action, understood as the process of change undergone by students during the learning process; and (5) evaluation, the awareness of both educators and students of such change, so as to make it grow.

Today, the new basis for educational contents and practices, all the more during and after lockdowns and distancing due to the pandemic, involves the metamorphosis of writing and reading into digital forms and courses delivery online. If the printing press instantly augmented and spread the distribution of linguistic content in space, digitalization has accelerated this distribution and access in time. To be rethought is not just the use of technologies but learning and teaching methods.

This has led de Kerckhove to the idea of connecting students in class as a paradigmatic pedagogical approach mirroring the networked model of the digital transformation. The principle is to improve learning and innovation through students collaborating and sharing. "Connected intelligence" (de Kerckhove 1997b) is neither "owned" by the single individuals, nor is it simply the sum of the links connecting them; rather, it is the outcome/surplus that derives from such rhizomatic connectivity. Students autonomously manage their research work in class (albeit supervised) and can continue it off/online. The method entails learning and problem-solving in small groups, based on a broad common theme, creating a synergic cross-fertilization. After deliberating on various subthemes of a specific problem, each group presents the finding to the class for feedback, with the goal to arrive at an integrated scenario made up of the contributions of the various groups.

These pedagogical methods—one posited on literacy, the other on associative thinking—need now to be reworked and adapted in sight of the emerging quantum ecology. The idea is to foster a *transdisciplinary method* that enables students to grasp and act upon the deeply uncertain and entangled reality. To be precise, multidisciplinarity is the extension of a given topic across disciplines and fields: it induces cross-fertilization osmotically; interdisciplinarity, instead, has to do with the application of concepts or methods across disciplines: the cross-fertilization is still induced, but it fosters stronger bonds that cut across human thinking.

Transdisciplinarity is something different altogether: it posits by default the unitary nature of knowledge and human thinking, and it aims to rework them from within. As Basarab Nicolescu (2005, 7) notes, "As the prefix 'trans' indicates, transdisciplinarity concerns that which is at once between the disciplines, across the different disciplines, and beyond all disciplines." More to the point, "transdisciplinarity concerns the dynamics engendered by the action of several levels of reality at once." Transdisciplinarity foresees an inherently holistic way to look and act in the world. To do this, transdisciplinary education is required, as one that is able to sustain such *modus cogitandi*. Nicolescu (2005, 9) argues that "the transdisciplinary viewpoint allows us to consider a multidimensional Reality, structured by multiple levels replacing the single-level, one-dimensional reality of classical thought." Significantly, by "classical thought," Nicolescu precisely refers to a condition based on classical physics, while he attaches transdisciplinarity to quantum physics. Following up on his reasoning, we could say, then, that while language and digital ecologies are part of the same level of reality—one based on classical physics—quantum ecology demands a different approach.

It will be increasingly relevant to teach not only skills and competences (sooner or later, with their takeover by generative AI, destined to deplete and become obsolete) but also methods and visions that cut through various disciplines. In a data environment where all answers are accessible and assembled on demand, students should be encouraged and learn to formulate questions. This "conversion" must be based, *ex principio*, on the cultivation of doubt not as an undermining skepticism (the all-encompassing conspiracy approach witnessed today) but as a cautionary, adaptive approach based on the awareness of the intrinsic ambivalence and openness of reality. This requires, in turn, a change of perception and action able to conceive phenomena as entanglements between observer–observed and one that is able to take a collective standpoint by default instead of particularizing learning. We envision topics such as "climate and society" or "civic technocultural education," which can synthesize lessons from various disciplines; we envision pedagogic modules focused, for instance, on "entanglement" or "uncertainty" as conceptual lenses for rethinking disciplinary boundaries; we envision whole curricula centered on "ecological thinking": at once a topic, a method of investigation, and a pedagogical approach; we envision the exploration of the double-sidedness of concepts and the unavoidable emergence of unintended consequences of tech implementation on the human–environment whole, as keys to disrupt problem-solving thinking in favor of *problem-opening* and *problem-seeking*.[6] Hence, transdisciplinary education overcomes axiomatic statements, getting rid of any privileged point of reference and rebuilding the co-subjective cultural fabric. Nicolescu (2005, 15) summarizes this very well, saying that "transdisciplinary education will allow us to establish links between persons, facts, images, representations, fields of knowledge and action . . . and to build beings in permanent questioning and permanent integration."

Sustainable Commoning Models

Over the past fifty years, a series of initiatives have been taken with the goal to rethink and rebalance current economic models of growth and development. These initiatives can be regarded as a long wave tide reverberating through the whole world with increasing force and urgency.

In 2012, the Rio+20 conference produced "The Future We Want," a document outlining a new globally orchestrated approach to global sustainable development. The work lasted over two years, involving actors from all spheres—economic, political, social, and environmental—and at all levels, from governmental to nongovernmental bodies, from authorities to citizens. The result was the publication of the *2030 Agenda for Sustainable Development*, which was unanimously approved in September 2015 by all UN members. The *2030 Agenda* defines seventeen ambitious Sustainable Development Goals (SDGs), subdivided in 169 targets, to be achieved by 2030, in order to guarantee long-term intergenerational sustainability. As Enrico Giovannini (2016, 41–42) notes,

> Three are the innovative characteristics of the *2030 Agenda*: 1) its universality; 2) the necessity that all contribute to change; 3) its integrated vision over the problems and actions to be realized. . . . There is the necessity to intervene in depth in how the world consumes and produces, creates jobs, assures the wellbeing of people, manages institutions, defines its value. . . . [We need] grassroots participation, which generates synergies, identifies innovative solutions, shares objectives and tools, checks results, and assesses the behavior of companies, politicians, media. . . . [The integrated vision] is the most difficult challenge, not only for its analytic complexity and practical feasibility, but also for the cultural context surrounding the *2030 Agenda*, almost "reluctant" to holistic thought.

Beyond the universality and collaborativity of the approach, what is worth underlining is the stress on the need for "holistic thought" able to produce an integrated action to socioeconomic–environmental concerns. While sustainability usually considers three main pillars—economic, social, and environmental (cf. Purvis et al. 2019)—already in 1987, the UN Brundtland Commission proposed, in response to the report "Limits to Growth," drafted by a pool of experts from the MIT, an approach also including the institutional dimension. In a similar vein, as director of OECD statistics, Giovannini supported the need to "move beyond GDP," identifying four forms of "capital"—economical, capital, human, and social—for the evaluation of sustainable development. Today, moves have also been made to include technology (George et al. 2020) and culture (Soini and Dessein 2016) as dimensions to be accounted in/for sustainability.

And yet, what if the problem lies not so much in extending the inclusiveness of the system under consideration, but more radically in its onto-epistemological foundations? Reasons are endogenous and exogenous. On the one hand, SDGs can be hardly considered as all equals. Nina Weitz and colleagues (2018, 531) note, for instance, that the *2030 Agenda*'s implementation is "complicated by the fact that targets and goals interact and impact each other in different ways." On the other hand, the tools and techniques used for monitoring the progress toward the achievement of these goals rest on forms of naive empiricism still based on "assumptions of objectivity, value-neutrality, and the ontological separation of subject and object" (Lake 1993, 405). Hence, a radically different approach is necessary.

To be sure, at stake is not a contrastive approach to development, such as "happy degrowth" (degrowth will likely be a consequence, including that of population, but it is not sufficient for the radical transformation the new paradigm demands) or "local resilience" (which, however effective it might be, can only go so far in rethinking society, since it works primarily as a lifeboat). These are valuable *tactics*, but they remain entrapped into a "productivity bias," whereby to be tackled are the externalities of people's actions on the whole system, not the reification that characterizes human relations to the environment in the first place. As Paul Shrivastava and colleagues (2020, 330) note, "In order to generate positive social and environmental changes globally, sustainability science must transform into a transdisciplinary enterprise."[7]

A more radical alternative is needed. An ecological *trans-action*. One that retrieves the etymological meaning of economy, that is, not the management of resources as it became known toward the end of the fifteen century but management of the house (*oikos*), namely the Earth. It is only from/ through such a perspective that it becomes possible to intervene into those "lock-in practices" that make paradigmatic changes difficult. To be discussed here is one example: commons-based peer-production.

*   *   *

"Modern industrial societies," it is possible to read on the P2P Foundation wiki page,[8] "are dominated by a materialist paradigm. What exists for modern consciousness is material physical reality, what matters in the economy is the production of material products, and the pursuit of happiness is in very strong ways related to the accumulation of goods for consumption." Such a materialist paradigm is, in fact, the by-product of a classical worldview that lends itself to mastering and commanding reality. Unless humanity is able to incentivize regenerative circular models, the current exploitative-by-default use of the environment (or human resources) cannot but have a deeply disruptive and socially unsustainable impact. Similarly, it is the role of technology as *dispositif*—not as a tool for exploitation, but as an organic element for integration—that needs to be rethought.[9]

Commons-based peer-production (CBPP) is a model that promotes the free engagement and cooperation of people to create shared value following community-defined governance mechanisms. Put differently, CBPP envisions forms of productive self-organization in parallel to the state and the market. In CBPP, "contributors create shared value through open contributory systems, govern the work through participatory practices, and create shared resources that can, in turn, be used in new iterations" (Bauwens et al. 2019, 6). Commons share and aggregate tangible and intangible assets in mainly four types: natural resources (including energy), labor, capital, and knowledge (including data). In CBPP, these aggregated assets are public and, being shared, are "owned" collectively so that individuals can freely access and use them, provided that the goal is not profit in itself, but rather the recirculation of the output (in any form this takes). Hence, the logic of privatized assets accumulation is marginalized in favor of the creation of a system where economic sustainability is to be achieved *alongside* social and environmental sustainability.

To favor this at both a local and global level—especially for the circulation of intangible assets such as knowledge and expertise—the internet as an open-access infrastructure is a key channel. Fundamentally, what technology can do must be evaluated in "humane" terms and not only at the service of humans. It is important to recognize the ecology in the making of which both people and technology are part: "humane," then, does not stand for "anthropological" but "sustainable." Here, the concept of "homeo-technologies" by Peter Sloterdijk (2000, 91) comes in handy: differently from "allo-technologies," which "violate" natural resources, "homeo-technologies are developed based on 'ecology' [and] entail a strategy of 'cooperation,' of 'dialogue' with nature."

Since collaboration is the *sine qua non* enabler supporting CBPP, this model foregrounds what could be called "peer interaggregation." Peer interaggregation can be understood as "commoning" (de Angelis 2017), that is, a sociotechnical process that repeatedly defines its relata, relations, and its own boundaries. The shift from commons to commoning is key to pass from things to processes. As de Angelis (2017, 11) notes, "Commons are not just resources held in common, or commonwealth, but social systems [of] ongoing interactions, phases of decision making and communal labor process." Commoning creates the commons as shared resources and values, not the other way around. Understood this way, peer interaggregation is nonappropriative by default (knowledge, technology, assets, and outputs are not owned, in the commercial sense of the term, but summoned up and recirculated), collaborative by design (it considers all nodes *and* edges of the system as integral and necessary to the system's flourishing), and collectively sustainable in its goals (indeed, commons for the community). CBPP, in other words, is one possible concretization of entangled social consciousness. As David Bollier and Silke Helfrich (2020, 5) note in *Free, Fair, and Alive: The Insurgent Power of the Commons*, "Any individual identity is always, also, part of collective identities that guide how a person thinks, behaves, and solves problems. . . . There is no such thing as an isolated 'I'. . . . Each of us is really a *nested-I*."

Talking of peer interaggregation also allows moving beyond an anthropocentric vision toward the conception of a system of peers where nodes (aggregates) not only are humans but also encompass organic and inorganic elements, as well as intangible assets. This is why the term *aggregate* fits particularly well: nodes are not fixed entities, but multifaceted instantiations that mutually reconfigure each other, in different scenarios and over time. It is self-organization through agreed mechanisms, rules, and roles on how to mobilize value, technology, people, and resources. When Floridi (2020, 120) writes that "every node-person is bound to all that already is, both *involuntarily* and *necessarily* and it should be so also *with care*," he underscores such ecosystemic view. Of course, there is no need to delimit this vision to people only, as Floridi later concedes: "The participation in the reality of whatever entity—including a human being—offers a right to existence and an invitation (not a duty) to respect other entities" (120). A quantum ecological framing helps to shift from a networked to an entangled mode of perception, so that Floridi's "invitation" eventually becomes a communitarian "necessity" engineered in the socioeconomic fabric.

In CBPP, the investment is made, from the outset, in the community as a living ecosystem of people, ideas, technology, knowledge, resources, and capital, rather than on the reification of relations and goals. A new holistic dynamic is being fostered. Two key terms need unpacking here: "community" and "general interest." A community is a fractal concept as far as its scale is concerned in that it depends on the interplay among, at least, three components: infrastructures (e.g., ICTs), law (e.g., national policies, regional directives), and locals' knowledge (e.g., people's practices, data, and relations relevant to and framed within a given place). As long as these three components are *ideally* coextensive (i.e., they overlap), action coincides with (and can be scrutinized in) the interest of the whole community. Whenever the coextensiveness of the three is not guaranteed, as it often happens, then self-organization fades, being substituted by top-down-only or global-market approaches.

This, in turn, implies that the general interest of a community is inevitably subjected to ongoing (re)negotiation. This is so because "general interest" is an entangled concept that demands constant contextualization. From an empirical perspective, the concept reflects the diversity of interests of all actors involved in a given situation; from an ethical perspective, it constitutes the synthesis (*not necessarily* the sum) of all actors' interests. In fact, such synthesis is never given once and for all; it is based on discontinuities across the community. Concretely, this demands the design of a participatory approach able to identify, negotiate, and adjudicate among such discontinuities over time.

It is not a matter of rethinking the relationship between the state and its citizens, or the role of the market, but of enacting a new sociopolitical-institutional design. Regardless of its form, such design demands the coupling of accountability and legitimacy not only as legal concepts, but as thick political practices in which the public—as a formalized sector and, above all, as collective assessment—regains a central role. Again, to adopt a quantum standpoint is meaningful as it recasts the argument under a completely different light: "a quantum social ontology suggests," according to Wendt (2015, 260), "that agents and social structures are mutually constitutive," that is, "a non-local synchronic state from which both are emergent." The ideas of "mutually constitutive" and "emergent" are crucial here: reality—even a socioeconomic-political one—is always in the making and its nature cannot be reduced to mere components or externalities. The individual is always already social, and any organizing form to come will require a set of checks and balances to accommodate that, if it really pretends to be sustainable.

The main problem that CBPP faces is scalability: while this model is based upon and promotes the global circulation of intangible assets, its concretization is inherently localized. Hence, making its model spreadable, scaling wide rather than scaling up, is an issue. However, such an issue is essentially methodological rather than substantial because it has not to do with the soundness of the model but with overcoming barriers that mainstream exploitative models still oppose to its deployment.

Quantum Communication Ecology

On the one hand, as seen in chapter 4, in the quantum ecology, human societies might witness the emergence of quantum sociotechnical world-sensing where the synthetic operating system of QITs creates macro techno-geopolitical filter bubbles. This in turn would produce enormous information asymmetries across different synthetic operating systems with mutually inaccessible filter bubbles based on their own self-validating self-legitimating logics. On the other hand, the world might witness an integrated synthesis on a world scale. The risk, in this case, would be a form of techno-homogenization that leaves no alternatives. To avoid the betrayal of an open, communitarian quantum ecology, it is necessary to avoid prescriptive enclosed scenarios and defend a wealth of possibilities, including the possibility of not being (technologically) determined. To do this, new ways of designing, managing, and operating QITs need to be envisioned.

First of all, it is necessary to recognize and accommodate ontological uncertainty into modeling and policymaking. It is not only a matter of minimizing risk (risk management); instead, it is the need to shift from a mastering to an adaptive approach to embodied knowledge and living. Martin Landau (1969), for instance, notes that the introduction of sufficient and appropriate redundancy makes any system "more reliable, more effective, more responsive, more able to withstand shock and damage than any of its parts." This view is similar to Elinor Ostrom's (2010) notion of "polycentricity," whose transposition on to the technological realm allows approaching technologies as *complex adaptive systems*.

From here, our suggestion is that QITs' management—their development, implementation, and use—should be one based on a republican approach. "To be a republican," Jamie Susskind (2022) notes, "is to regard the central problem of politics as the concentration of unaccountable power." Therefore, a quantum republic envisions roles, rules, and mechanisms to keep the whole ecology in balance, involving multiple stakeholders (including nonexperts), while also remaining open to adaptation to changes (Calzati and van Loenen 2023a). To design a republican approach for QITs means to realize a decentralization of the control of these technologies and a systemic distribution of the oversight processes concerning their implementation (and possibly development), to make them as resilient as possible. As researchers we can, first, denounce and try to avoid the mistakes that have been done concerning the co-optation of the Web by big tech companies: should these mistakes repeat with QITs, consequences would be much deeper; second, we can call for an international agenda that outlines and defends a republican approach to QITs at different scales and in different contexts—along the line of the dual governance discussed in chapter 4—so as to prevent the polarized scenarios above. Third, we advocate the need to develop quantum literacies through the design of transdisciplinary curricula tailored to different audiences.

*   *   *

When combining quantum computing with quantum communication, the result will be fully sealed "blind" computation: "nobody in the network can intercept data, not even the people who own the quantum computer."[10] Moreover, any node of the quantum network can be either source or destination, without fixed, preestablished order (Possati 2023), implying forms of communication that are, at once, more flexible and more unstable (Illiano et al. 2022).

Quantum teleportation is one of the most promising areas of research in quantum communication. Basically, quantum teleportation is a protocol for transferring quantum information (qubits). Differently from analogue or digital communication, quantum teleportation does not imply any physical transmission (for instance, of a text or bits) but instantaneous transfer of quantum *states* (for a critique of the notion of "state", cf. Pusey et al. 2012). A simple scenario includes a sender, a qubit of which one wants to teleport the state, an entangled quantum state, a traditional channel, a quantum channel, and a receiver. Sender and receiver can be any human or artificial nodes; the qubit is a two-state quantum system (which, being in a superposition, cannot be simply copied and sent—that is why quantum teleportation is key); an entangled quantum state is, for instance, two entangled qubits (in four possible states); the quantum channel is a communication channel for (simultaneously) transferring quantum states; and the traditional channel is a channel permitting (classical) communication (in this case of the measurement of the qubits, thus quantum teleportation can never occur at faster-than-light speed). Procedure: an entangled quantum state is created and then distributed to two different locations, A and B; at location A, the qubit to be teleported and one qubit part of the entanglement state are manipulated (quantum logic gates are applied). Then a measurement is conducted (and the initial states of these qubits are destroyed). The measurement triggers an instantaneous determination of the entangled state of the qubit at location B. The result obtained at location A is classically sent to location B, where the initial state of the qubit can be re-created depending upon the measurement result.

Quantum teleportation can be best grasped through math formalism, which allows to express entanglement and quantum gates in formulas, while language necessarily imposes a phenomenological sequentiality of meaning-making that cannot fully render the simultaneity that teleportation entails. For the sake of understanding, a graspable everyday experience might be the following: suppose that Alice and Bob share a manuscript (the piece of information to be teleported) on a cloud server (quantum channel), but they have not viewed it yet. First, Alice accesses the document and makes some changes to it (what equals to "measurement" above). These instantaneously change the document that Bob will view on his laptop (which is "entangled," being shared on the cloud). When she is done, Alice contacts Bob to tell him that she has finished implementing changes and asks him to have a look at the document. So, Bob accesses the document, reads it, but then also wants to retrieve the original version of the document to compare the old and new one. What he does is to go through the revisions' history (i.e., which amounts to a classical channel) and cross-check the changes made by Alice, eventually reconstructing the original manuscript.

From a classical media perspective, one might say that quantum information is the message (to be transferred); entanglement is a form of *shared metamessage* associated with the message; measurement is the process of signification allowing for disentangling the message; quantum channel is a

metamedium, while classical channel is a traditional medium (for communicating the result of the signification process). What is instantaneously transferred through quantum teleportation is a double "message + metamessage"
that gets disentangled through signification/measurement; then, interpretative rules apply at destination to the metamedium-teleported "message +
metamessage," allowing for recoding the initial message (thus, interpretative rules classically explicate the process of signification). As Illiano and colleagues (2022, 14) write, "As long as an entangled state is shared between two
nodes, they can transmit a qubit regardless of the instantaneous conditions
of the underlying physical quantum channel."

In this understanding, the traditional bond between medium and
message—"the medium is the message"—is reworked; through entanglement
as a shared metamessage, it is the process of signification that dictates when
and how a message comes to be. Hence, the traditional "sense" of communication—in its *three-folded connotation*—gets inevitably affected across
nonlocality and temporalities. This opens the door to new ways of thinking about communication, moving beyond a theoretical framing that links
agents, messages, and media under a classical triadic scheme; by contrast,
the signification process at the basis of the scheme remixes and synthesizes
sense as a self-standing unit, with consequences on the architecture supporting referentiality, trust, interpretation, and access itself to communication.

Urban Ontologies as Caring Urban Environments

In chapter 3, we have discussed smart cities and city digital twins as examples of the technohuman entanglement produced by digital transformation
in an urban scenario. We have also mapped the epistemological tensions
that lie beneath these concepts and the socioeconomic–environmental
consequences of their implementation. Here we discuss one possible way
for inscribing the city into an ecological vision that regards the tech-urban
environment as a complex evolving whole.

Hence, we oppose current "control urban spaces" (CUSs)—that is, how
diffused technologies and digital models in/of cities are largely conceived
nowadays—to the idea of "caring urban environments" (CUEs), in which
technology becomes a *homeo*-means (see above) for addressing macro
socioeconomic–environmental urban dynamics in sustainable ways. In other
words, CUEs are *emerging sociotechnical environments taking care of themselves*,
that is, being constantly reworked from within through forms of ecological

self-organization. Hence, a whole new being-politics comes to life in which it is the whole (e.g., carbon footprint, data–people symbiotic trade-offs) to dictate its own conditions of existence. At the core of this new being-politics, as a biocultural embodiment, is design, intended as an endless reconfiguration of possibilities, rather than as a solution-oriented approach. In this, we follow Easterling (2021, 12) when she states that "design is *indeterminate in order to be practical* [italics added]." Design is a door kept open, a strategy that cuts through different ecologies, a quest for ever new entanglements between the physical, the cognitive, and the lived spaces, in view of sustainability.

Concretely, to the extent to which, as seen, QITs will redefine traditional categories of individual and collective responsibility, space, time, accountability, transparency, and trust, CUEs by and through these techs shall be designed as a *commoning practice* that is (1) context based, (2) iterative, and (3) participatory, in order to make these CUEs truly ecological. First, CUEs require the cognizant study of the urban environment, including current and past socioeconomic–environmental dynamics that make such an environment unique. This contextuality can also be projected on the commons as resources and linked to an issue of value: Which values does a certain community prioritize? How has it arrived to do so? Why? How can homeo-technologies align to that? Only a contextualized transdisciplinary analysis can provide pertinent answers, helping design a fair integration between data, people, and context.

Second, CUEs must be able to perform iteratively, adapting to change and circumstances. This means that the commoning needs to mitigate potential adverse events, as well as to accommodate unforeseen inputs. As a complex ecosystem, the urban environment configures an interacting—with both biological and nonbiological elements—and intra-acting—within itself—dimension, and this dimension demands to be kept in balance over time. Hence, CUEs must continuously account for (in and out) relevant elements, while striving for an overall equilibrium. When de Angelis (2017, 17) writes that "the subjects of this movement, the commons, are not here understood as individual subjects, but as already systemic subjects," he points exactly to the codependency between individual and collective stances and to commoning as a practice that negotiates between these two. The process is iterative to the extent it must allow for a moment of "fixation" of the commoning process (e.g., data and practices) while also collectively producing its own assessment.

Third, CUEs must be participatory. Participation can be disentangled according to three axes: *Who* participates? What *kind* of participation is at stake? *How* is participation designed? Concerning the first question, by now the quadruple helix approach—involving private actors, the public sector, academia, and citizens—has become the standard to achieve an inclusive urban development process. Yet, from an ecological perspective, the quadruple helix is not enough; it should be better regarded as the baseline, rather than the optimum. In fact, a whole galaxy of noninstitutional actors does contribute to inform the development of the city: NGOs, nonprofit organizations, intermediaries, citizen communities, and so on. Only fair mechanisms of in/exclusions on a rolling basis of all these actors can provide a robust framework for governing such a heterogeneous galaxy.

This leads to the second question, that is, the kind of participation at stake. Sherry Arnstein (1969) developed a ladder for evaluating the level of citizens' participation in the public sector, identifying eight steps over three degrees of participation: "nonparticipation," "tokenism," and "citizen power." According to Arnstein, it is only in the last three steps at the top of the ladder that citizens are really empowered, having effective and direct accountability and deliberative powers over the decisions to be taken. To have a successful participation, then, participants need to clearly know which goals they want to achieve and how, that is, to "manage the system as a process of continuous innovation, learning and adaptation" (Toots 2019).

This addresses the last question on how to design participation. In this respect, participation needs to be regarded as *open-ended*, that is, one that admits conflict and dissonance, as well as multiplicity of the "possible," as long as this multiplicity does not affect collective interest (cf. Sennett 2018). Participation, then, demands as much contextual iteration as synthesis among different views; at the same time, participation is inclusive in the sense of plural and yet always incomplete, possibly irreconcilable. Similarly, commoning is entangled with participation to the extent to which participation sets the (ever-temporary) boundaries of systemic autonomy as a self-organizing entity.

Literature, Cinema, and the Arts

We have seen it at the beginning of this book: quantum concepts tend to be transposed onto other disciplines due (also) to the fact that they are intrinsically transdisciplinary. As a meaning-making *dispositif*, the language used to describe quantum physics is inherently heterodox, heretic, in that it is entrusted with a task it cannot fully accomplish, given that quantum physics transcends phenomenological intuition. The more humans dig into physical reality, the more language becomes an approximation, insofar as the gap between percepts and concepts widens. To an extent—and quite ironically—this discrepancy has already been put *into words*. As Ludwig Wittgenstein (2016) wittingly wrote, "Funny that in ordinary life, we never feel that we have to resign ourselves to something by using ordinary language," implying that language enforces the shaping of a certain "reality" to the detriment of potential others.

Artistic forms impact on humans' understanding of the world by creating connections where there seem to be none; they show perceptual–cognitive arrangements never suspected before. This is why art is able to shift traditional worldviews: it forces to rethink the meaning-making of what constitutes reality. An art that reconciles object and subject, interpretation and presence, text and body, meaning and experience cannot but be ecological, reworking reality from within, being able to show and (make) feel the presence of the unexpected, or the hardly suspected. As Guattari (1995) notes, "An ecology of the virtual is just as pressing as ecologies of the visible world. . . . In this respect it is not only fractal geometry that must be evoked, but fractal ontology." Examples abound, from *sur*-realist paintings, to Cubism, and Futurism's poetry—which indeed instrumentalized language—all the way to John Cage's music as a remolding of sound beyond established perceptual and cognitive patterns.

Sarraute (1939), a *new novelist* Russian-born French author, published in 1939 a series of short stories, or observations really, that she called "tropisms."[11] Tropisms are liminal mental sensations at the border of awareness, which are by default co-subjective, bearing the mark of the context with which they are entangled: "[Tropisms] are indefinable moments that slip very quickly to the limits of our consciousness; they are at the origin of our gestures, our words, the feelings that we manifest, that we believe we are experiencing." As an avant-gardist, Sarraute tried to reinvent literature by working through language, and most interestingly, she did so influenced by the physics' discoveries of her time, as she acknowledged. Sarraute's experimentations point to a hermeneutical question that is worth asking: Is it possible to rethink language beyond/through its representationalist function? What could this language be? For instance, which shape could an indeterminate and/or entangled quantum writing take?

In a seminal group of essays under the title of *L'ère du soupçon* (*The Age of Suspicion*, a title oddly reminiscent of the uncertainty principle), Sarraute (1956) prefigured what would become the short-lived but potent decade or so of the *nouveau roman* bringing together novelists who put into question time, space, and self in an original manner. To be resisted first of all was the narrative, which became open-ended, as per the widely successful essay by Umberto Eco (1962), *Opera aperta*. Narratives become tentative, exploring possibilities but arriving at no definite conclusion. Most likely under the influence of cinema, the one-way forward time sequence of the traditional novel gives way to a continuous present tense; perhaps the most deconstructed feature of conventional literature is that of the protagonists, rarely named (as in most works by Sarraute, Alain Robbe-Grillet, Michel Butor, Nobel laureate Claude Simon, Marguerite Duras, and many others) and more often than not hardly situated. A startling example, Butor's (1957) *La modification* is written in the present tense and the second person to involve the reader on a train ride between Paris and Rome in changing his mind about divorcing his wife and living with his Italian mistress. Examples all seem to indicate a kind of restlessness in the *Zeitgeist* that could be distantly attributed to a background change in physics, but more directly to the dislocating effect of film editing that "cuts" into time and space and summons a fully experienced reality much in the way Barad (2007) describes intra-action.

Indeed, despite its intellectually and epistemological transforming effect, or maybe because of it, the *nouveau roman* never made it to the mainstream and to this day remains a distant literary cult genre not much emulated. What novels could never resolve but only allude to cinema extended to the whole world's population almost overnight, which is to make anybody who ever entered a cinema or watch a film experience space, time, and self as never before, and this without considering quantum physics for a nanosecond. To wit, time edited in parallel with the continuity of the spectator's duration, time accelerated, reversed, slowed down, time modified, contradicted or expanded by musical effects, voiceover and other cinematic wizardry; space repeated, repositioned, dislocated, reoriented; the self's point of view substituted, divided, or redoubled by that of the camera; and above all, a world summoned instantly by "cuts."

Among the uncanny "effects preceding causes" dear to McLuhan is a 1946 book, *L'Intelligence d'une machine*, by French filmmaker and critic Jean Epstein ([1946] 2014, 125), who suggests that "the cinematograph rather abruptly ushers us into the unreality of space-time." Avowedly inspired by contemporary discussion of quantum physics and writing about "Heisenberg's inequalities" (97), Epstein dismisses the same dichotomies that preoccupy today's critical theory and philosophy: "Very old, perennial problems—antagonisms between matter and mind, continuity and discontinuity, movement and stasis, or the nature of space and time, and the existence and inexistence of any reality—come into view under a brand new light" (xi). Observing the cinematic accelerated growth of a plant, Epstein concludes that "time is not made of time" but "merely a perspective resulting from the succession of phenomena, the way space is merely a perspective of the coexistence of things" (21). Emphasizing the correspondence between Epstein and Barad's terminology and thinking, Christine Reeh Peters (2023) notes,

> What does it mean to be "intra-actively produced in the making of phenomena"? How can the cinematograph sense reality by simultaneously producing mechanical thought? Epstein writes: "It is not so much that humans or their machines discover a reality that would preexist, but rather one that they construct according to the pre-established mathematical and mechanical rules of space-time. Reality, the only knowable reality, does not exist: it is manufactured, or more precisely, it must be manufactured. . . . This is only possible through the pre-conceived framework determined by the constitution of the operator that activates the formula, that is, through the thinking apparatus, whether it be human or inhuman."

Pushing the comparison further, Reeh Peters concludes, referring again to Epstein:

> The recorded worlds thereby manifest themselves in the form of technical images and sounds, because this is the condition of the measuring instrument, the cinematograph. The images and sounds are not properties of the human mind, but, on the contrary, point to an intra-active relation of matter and meaning reconfigured by cinematographic thought, understood as the topological machine intelligence "inherent to the most advanced mathematics of realism."

Thus, the idea that gives its title to Epstein's book has to be taken literally. For him, cinema is a manifestation of the intelligence of a machine, the actuation of thinking by a machine, that is, an apparatus that does not only refer to the kinds of images and sounds that one would construe mentally from written words, but actually produces them. A thoughtful critic of cinema, Benjamin (1999) considered the seventh art as a technology capable of introducing the subject to an "optical unconscious," that is, to details not perceivable by the human eye, which then contributed to alter the relation of the subject with reality. Across various of his writings, Benjamin argues that in and through technology, a new *physis* is being organized, meaning by that the organization of a "collective body" in which humans and technology are mutually integrated.

Without espousing transcendentalism, here we sympathize with a certain critique against interpretation, and yet, not really in opposition to experience but from an ecological standpoint, always aware that there is no privileged point of reference from which to judge, but only an implicate order of signification. As Simanowski (2011, 210) argues, art forces a hermeneutic enterprise that leads "to experience the relativity, uncertainty, and infinity of signification." This is a quintessential characterization of the hermeneutics of the quantum ecology.

In this respect, Hans Ulrich Gumbrecht's (2012) idea of *Stimmung*—translatable as "atmosphere"—hits the target. While Gumbrecht focuses primarily on literary works, his idea is extendable to other semiotic processes, in that *Stimmung* broadly describes the *Geist* of an artwork, intended as an experience that "envelop[s] us and our bodies as a physical reality; something that can catalyze inner feelings without matters of representation necessarily being involved" (5). From an onto-epistemological point of view, Gumbrecht's idea implies that, in order to account for people's being-in-the-world, language alone is not enough; it is crucial to cut through and past language.

To do so, the palette of signification must be expanded in terms of not only forms and contents but also cognition and apperception, that is, being itself *as* technology/*dispositif*: in Guattari's words (1995, 89), "The extraction of deterritorialized percepts and affects from banal perceptions and states of mind takes us from the voice of interior discourse and from self-presence on paths leading to radically mutant forms of subjectivity." Note that this deterritorialization is already happening with AI increasingly put to use to decode animals' expressions, as well as plants' emitted frequencies eluding human senses. At stake is not (only) a coarse issue of translation but the reconfiguration of a whole ecological horizon of signification, or world-sensing in its truest meaning. It is still too early to predict the impact of this human–environment–technology symbiosis, but for this very reason, a space for collective resensing of reality is possible. Which role might QITs play in that? Artist and physicist Libby Heaney has made this question the premise of her artistic works and installations, among which *Ent* (2022) allows users to "navigate non-binary landscapes & encounter entangled forms & quantum generated hybrid creatures,"[12] while *Q is for Climate?* (2023) is an immersive exhibition exploring how people can learn to think and feel like the climate itself, also questioning the upcoming impact of quantum computers on the environment.

With regard to her book *Entangled: Physics and the Artistic Imagination*, which contains interviews with physicists and artists, Ariane Koek (2020) notes that these testimonies "reveal the differences as well as connections of seeing the same phenomena through different eyes. It is my belief that it is in the connections, the differences, and the gaps in between, that the two most unique aspects of being human thrive and grow—the sacred space of the imagination and creativity." This is exactly the kind of synergies that are increasingly relevant for shaping new interpretative frameworks, means, and courses of action, in order to make sense and act within the emerging quantum ecology. Most important, it is necessary to refract these synergies back onto the language(s)—here, in the broadest sense—adopted to express them, so that humans' grasping of reality can get enriched and diversified.

Sailing toward a New Paradigm

Every action already contains its own reactions (plural), and it is not possible to ever get a full mapping of the scenario at stake. This is what an onto-epistemological reading of reality by quantum physics forces us to accept and one that hosts our idea of *dispositif*-dependent open reality. Calling for an ethical responsibilization is not sufficient: on the table is something more and different. Notably, it is the dethroning of the "human" from its self-imposed privileged position within the world. A shift—based upon a new vision, surely but not only, a new feeling too—is needed if we want to survive and thrive as a collective.

For millennia, humans have molded and remolded "reality," capitalizing on their capacity to represent and meta-reflect upon their own condition through different language-based operating systems. Since humanity has records, people have sought Truth, Beauty, God, Universal Harmony, and Laws, and they have done so, in the first place, because they were able to cooperate and communicate, sharing anxieties and desires for knowledge. Language has provided the channel *and* stuff for (a certain) "reality" to come to life. This practice has produced beauties, truths, gods, harmonies, and laws of various sorts, but, most important, it has triggered new questions, new anxieties, and desires, widening rather than reducing the scope of the unknown. The other side of language is—quite evidently—the reification of *techne* into a techno-logos (i.e., a disciplinary enframing of the human capacity to interface with and shape physical reality). Language and technology are inseparable, both bearing the mark of the ever-approximate mastering of world-sensing they can supply. It appears under a particular light, then, the teaching of Laozi, according to which "the Tao that can be told is not the eternal Tao." Every *dispositif* produces one possible "reality," /reality/, {reality}, or else.

The coupling between language and technology has been harmless—at least from a sustainability point of view—until technological development has reached a critical point at the intersection of four tendencies: population growth, natural resource exploitation, unequal redistribution of wealth, and the ultimate possibility of humans' self-destruction. At stake is a techno-linguistic-economic bundling so tight that it often appears as an unquestionable condition. It is not a minor historical irony, as Benjamin Labatout (2021) acutely describes, that Fritz Haber, the German scientist who invented modern fertilizers at the beginning of the twentieth century—thus favoring an exponential demographic growth worldwide—was also the man who invented the first chemical weapon of mass destruction, used since World War I.

It took two world wars for humanity to start to question the fragile and increasingly dangerous bond between language and technological development. In the mid-twentieth century, the world sought a new order, and the internet has materialized this new desire in the form of a global rhizome. Hierarchies did not disappear (as power struggles did not), but (societal) pyramidal structures were gradually flattened under historical pushes, opening the way to collaborative connected practices. These practices, in turn, found in the synaptic circuits of the brain a validating metaphor: language sewed the coupling of human and technology, but in this case, it was the whole human to become a technology. The equation of the brain as a wired machine created the needed cognitive space for the nascent AI research. Eventually, the internet did really rewire human thought and experience, maybe not as optimistically as hoped at the beginning, but certainly in a way that forced the whole world, for the first time in history, on the same sociotechnical page. It linked—and coalesced—the village and the globe and arranged all its actors in the present, however virtual that dimension might be. And yet, it was illusory to expect a wholly interconnected globe as fully harmonic, especially to the extent technology did come with a price tag: economic inequality and environmental exhaustion.

We do not seek historical accuracy here, but the most dramatic turning points that stand for the ethos of entire sociotechnical epochs. Suffices to think that already at the time of the internet, the third ecology—the quantum one—was already in the making and concretized in technological developments that further fueled the digital ecology (e.g., GPS). Hence, if it has been for decades, if not centuries, that humanity has felt in a state of constant crisis, that is because technological advancements have increasingly widened the gap with language in delivering a coherent world-sensing. Technology has developed too quickly, not giving people—their body–mind affordances—the necessary means to adapt. The present has become a crumbling time, eroded by its own techno-determined anxiety. The crisis of the Anthropocene is an entangled condition in which language seems to have exhausted its resources and technological development to have slipped out of hand. And yet, crises are also moments of reflection and change. And we do hope that this book can contribute to both, working as a lens *and* a prompt to action, a framework and a case study. In fact, the book introduces a varied array of doubts and questions that beg for further investigation: How can quantum politics be performed? What quantum culture(s) will emerge? How to provide people with the tools and strategies to act for good in the new ecology? And what about issues of literacy, civism, self-determination, sovereignty, and activism within such ecology? To what extent are these concepts valid and how should they be questioned and thickened beyond the normative (based on classical physics) worldview?

A core argument of this book is that with the advent of QITs, the tensions of the "Anthrobscene," as Jussi Parikka (2014) calls it, will become even more evident and radical—and the exacerbation of conflicts around the world is just another sign. More evident because they will impact by default on the whole globe, and this will happen starting from an already multipolarized condition (differently from the tabula rasa produced by the H-bomb); more radical because the onto-epistemological tenets at the basis of quantum physics are different from the classical ones on which both language and digital ecologies rest. Uncertainty and entanglement will become crucial lenses through which to interpret and reshape reality, while the upcoming synthetic operating system will remold it from within. The linguistic and the networked paradigms will certainly not be dismissed, but they will be integrated into a new sociotechnical scenario that demands humans to come to terms with their enmeshment in the cosmos and the intrinsic partiality of all knowledge.

Ecological disasters such as climate change and the Covid-19 pandemic, as well as those brought by wars, manifest the misalignment between people being, knowing, and acting in the world. It is time for a change. In this last chapter, we outlined some major criticalities and possible alternatives. But the road ahead is still long. Regardless of the various ramifications that the quantum ecology can take, its shaping is a political choice. Politics is not some abstract concept; it is a praxis traversing all sectors and fields. Politics is a unifying field in which all should have the chance to be involved not just as individuals but as entangled aggregations.

In this scenario, education is key, but so is the disjoining of technology from mere profit logic. Today's economy materializes *do ut des* power relations, oblivious of externalities or nonquantifiable values, especially in the long run. Politics must undo and redo power relations keeping in mind a sustainable perspective. The reason is that this will have social and environmental, *as well as* economic, benefits. But this can only pass through new means of expressions, new epistemologies, which quantum physics can help sustain.

It is not difficult to spot the elephants in the room: despite valuable initiatives (e.g., the UN *2030 Agenda*), actors around the world—the EU, the US, China, but also India and Brazil, as well as the whole African continent—can no longer postpone their actions. The *destruens* path humanity is on needs to be reversed into a *construens* path, and this inevitably means to embrace a different approach to the world. Nobody, for better or worse, can be excluded. In times of increasing transparency (public and less positively private), the system can no longer tolerate the obdurate opacity of big tech giants and of political powers. Or else, it is not difficult to predict that disruptive events will repeat in different forms and in ever-deeper acuity. *Unless* this is what economic centers of power want for the majority of humanity (but then, of course, a whole different reflection follows). This is the bad news.

The good news is that in this transition, there is hope. And this hope comes, above all, from new generations. The example of the "Fridays for Future" international mobilization is just a case in point. Different from other mobilizations, Fridays for Future is sustained by an aggregating message. Indeed, nothing is more collective than the environment we live in. It concerns us all. Moreover, it is a peaceful mobilization, and it is led by those same young people that adults have for too long and wrongly assumed to be drowned into the digital ecology. In fact, these young people have shown the willingness, even the eagerness, to have their voice heard in the physical reality, in the streets. And they rightly do so, given that the world will be much more in their hands than in those of older generations. The duty of these latter is to support the former: their success is the success of humanity.

Afterword

Derrick de Kerckhove

Old men ought to be explorers
Here or there does not matter
We must be still and still moving
Into another intensity
For a further union, a deeper communion.
—*Four Quartets*, Thomas Stearns Eliot (1940)

La vida es sueño
—Pedro Calderon de la Barca (1635)

History is a nightmare from which I am trying to awake.
—*Ulysses*, James Joyce (1920)

Un rien imperceptible et tout est déplacé.
—*Le Maître de Santiago*, Henri de Montherlant (1947)

Collaborating on this book has been an unusual adventure. Traveling to understand quantum physics from a nonexpert approach, we ended up, literally, at the edge of reality. Having been there before but without the benefit of a scientific investigation, I could recognize it.

Literature, Western, Eastern, or even Aboriginal, contains a wealth of suggestions to support quantum physics' basic tenets. Joyce's *Finnegans Wake* is a pun-filled display of superposition and uncertainty to educate *abcedminded* people about the blinding power of literacy from *alphabetters* to *alforabit*, his lightning insight about the coming digital transformation. But Joyce's most telling line is that of *Ulysses* quoted above. The suggestion that reality is a dream is not new as it is found from Heraclitus and Plato all the way to Joyce via Zhuangzi and Descartes. What is surprising is that in all such elucubrations, while most mention the fact that the dreamer is not aware of only dreaming, none questions the experience of the dreaming self occupying centrally the scene of the dream.

Letting physicists, "realists" or otherwise, argue about the objectivity of the real and/or its dis/continuity with the observer, an interesting—and promising—approach is to suspend belief about the continuity of existence and imagine, better, experience, the appearance of the world as an epiphany, as something that has "just arrived" and that keeps arriving for the first time, every time people apply a critical awareness to their live participation in the world instead of taking it for granted. Of course, this alternate response to reality is, or rather was, counterintuitive before quantum physics introduced a serious doubt about continuity. The perdurance of their surroundings and the seamless continuation of their lives, plus or minus moments of inattention or forgetfulness, would naturally lead people to trust absolutely that their house or their office has not moved, changed, or disappeared when they are traveling elsewhere. Cats and dogs too know their way back home without trying.

Experiencing the world as a sudden epiphany instead of an objective continuum is not as difficult as it may appear at first. We do it every night, at least those among us who dream. During the time I dream, I am participating into a "real" experience and, only in retrospect, when I wake up do I realize that not only none of it happened but that I ruefully attributed to places and events the instant recognition of extant reality and referentiality despite the same places and events being substantially different—some even nonexistent—from the ones that I really know or remember. To say nothing, of course, about the defunct people who somehow "resurrected" without doubt or question in my dream.

The other thing to note about the similarity between dreaming and the world as epiphany is that both depend upon the specificity of my unique and central presence in the scene. I, the dreamer, have created the scene, with my personality, my previous experience and even, in turn, depending on where I am living at the time, using the three languages I speak. Now, while my dreaming consciousness has clearly summoned the experience, one could object that here, as I write, my wakeful condition differs absolutely from the dream. How could I pretend that it is me who summons the reality of the world out there with all its complexity, cosmic dimensions, and accumulated memory? Precisely because of entanglement. What is common to both dream and reality is that both are totally entangled. I am no more in the world than I am in the dream; I am an integral part and parcel of both. What distinguishes both experiences is that the first is shared, but not the dream.

Precisely because it is shared, the entanglement that binds me to the world brings on a kind of responsibility that is not in a dream (apart from wakeful or "lucid" dreaming, a different story altogether because while there is a sort of responsibility, it is only to a fiction about which the lucid dreamer is fully aware). In the world, I have power of action and decision that are effective, even at the slightest move. To be or not to be is to be responsible, if not to others, at least to oneself as we propose in chapter 1, or in a quantum world-sensing to the whole environment. Different religions and secular ideologies have primarily oriented and justified the kind and the object of responsibility individuals and societies should take on and respect each according to their views. They may have worked for a while, but it would be difficult to assert today in view of the disastrous world scene that they are doing very well at all. But then, why would reconsidering the world as an epiphany in which humans play an integral, participative, and ever active part be different?

It would be different because, by becoming aware of being co-substantial and bound to the whole, of which each one is a unique part, would change the ground of existence itself. The axis of responsibility would not be to the other as in "shame" cultures, or to the self, as in "guilt" cultures, but to the world, beginning with the Earth, of course, as the ancestors, the "primitives," the Aboriginals saw it. With that in mind, would society continue to tolerate the suffering, not only of people abjectly left to burn, drown, and starve under climatic conditions that greed created, but also of butchered animals and torn plants and mined rocks denied the respect they once received from people less technology-savvy than today?

"Technology," said to me an Aboriginal participant in a Connected Intelligence Workshop in Melbourne in 1995, "is a very recent thing. We Aboriginals have survived 50,000 years without technology." Anthropology tells us that Aboriginals have no word for time, nor did they have a concept for it until colonization. I would be tempted, of course, to continue my thought experiment above on the much-bandied notion of "dreaming" and "dreamtime" in Aboriginal culture. The term, however, is strenuously disputed among anthropologists, and so is the romantic interpretation to imagine that all nine hundred different communities of Australian Aboriginals practice dreaming as a duty to keep reality going. On the other hand, looking into the matter of spiritual experience shared by most if not all of them reveals that these communities are perhaps even closer to a quantum-inspired sensibility than what we learned and wrote about traditional Asian concepts:

> Our spirituality is a oneness and an interconnectedness with all that lives and breathes, even with all that does not live or breathe. . . . Everything else is secondary. (Mudrooroo 1995, 43)
>
> It is a principle of connectedness that underpins Aboriginal life. And because of connection, [the practice of] Kanyini teaches to look away from oneself and towards community. (Randall 2003, 12)
>
> We do not separate the material world of objects we see around us with our ordinary eyes, and the sacred world of creative energy that we can learn to see with our inner eye. (Randall 2003, 16)
>
> We work through "feeling," what white people call intuitive awareness. . . . White people separate things out, even the relationship between their minds and their bodies, but especially between themselves and other people and nature . . . [and] spirit. (Randall 2003, 24)

When QITs—and attending epistemology—come to fruition, they will have to address all of nature, not just the fate of humans. They shall pay the kind of ancestral respect that our forefathers had for plants, for animals, for hills and rocks and waterways. Paradise regained, you say? Maybe, maybe not, but at least, two master guiding principles issuing from understanding what quantum physics implies are now available and eminently shareable to chart the way ahead, *entanglement* that puts the whole world and cosmos in an undividable unity, and *uncertainty*, the key to change, without which nothing moves. Uncertainty is where humanity stands now.

Notes

Introduction

1. *Quantum technology* is a broad term that we use for simplicity to refer to technologies that rely on quantum mechanics behaviors and principles. In chapter 4, we will unpack this broad term distinguishing between *quantum sensing technologies* (already in use) and *quantum information technologies* (emerging as we write).

2. A terminological clarification: in the book, *quantum mechanics* and *quantum physics* will be used interchangeably.

3. In this context, *digital transformation* and the associated idea of *digitalization* are used as synonyms to refer to the digit-based procedures that "translate" the physical into binary code. This translation is an ongoing process involving organizational, technical, political, and socio-cultural-psychological-ethical effects (differently from *digitization*, which refers more strictly to the transposition of an analogically written text into a digital format, thus configuring a procedure that, while having certain effects in terms of production and fruition, has an identifiable beginning and end).

4. In doing so, we regard the Copenhagen interpretation, which was formalized in the first quarter of the twentieth century, as the scientific yardstick. While this interpretation has undergone updates and revisions, it is today the most accepted—although nonunanimously—framework for making sense of how the subatomic world behaves. The Copenhagen interpretation sprung out of the passionate debates originating across a heterogeneous group of physicists, including, among others, Werner Heisenberg, Erwin Schrödinger, Albert Einstein, Paul Dirac, Wolfgang Pauli, Max Born, Louis de Broglie, and Niels Bohr.

5. A cloud chamber is a particle detector that allows visualizing the passage of ionizing radiation.

6. Hugh Everett (1957) radicalized Bohr's position, quantizing the act of observation, which led to avoid the collapse of the wave function in the many-worlds interpretation.

7. Since 2021, measurements on the decay of muons (a charged lepton) have provided unexpected results based on what the Standard Model predicts, leading to suggest the presence of a fifth force.

8. According to Hertog, holographic physics might "complete the unification of general relativity and quantum theory that string theory had initiated. . . . It showed that gravity and quantum theory need not be water and fire but can be like yin and yang, two very different yet complementary descriptions of one and the same physical reality." This same reality can be thought of by recurring to the idea that the four-dimensional world of everyday life is a hologram, that is, information encoding flattened on a two-dimensional surface.

9. *The Economist*, April 28, 2023, https://www.economist.com/by-invitation/2023 /04/28/yuval-noah-harari-argues-that-ai-has-hacked-the-operating-system-of -human-civilisation?utm_content=article-link-3&etear=nl_today_3&utm_campaign =a.the-economist-today&utm_medium=email.internal-newsletter.np&utm_source =salesforce-marketing-cloud&utm_term=4/28/2023&utm_id=1579250.

10. This is why it is not possible, yet, to speak of quantum "innovation" and apply a technoeconomic lens (cf. Perez 2010). This, however, should not dispense from envisioning ahead what the quantum ecology will look like.

11. As a concept, the "point-of-being" aims to legitimize an embodied sensation of the world and a resensorialization of the environment to complement the visually biased perspective with a renewed sense of humans' relationship to their spatial and material surroundings. As such, the concept entails a topological reunion of sensation and cognition, of sense and sensibility, and of body, self, and world, displacing the self through the physical, digital, and electronic domains that shape today's physical, social, cultural, economic, and spiritual conditions.

Chapter 1

1. Interestingly, this hiatus also marked, according to Hertog (2023, 80), the work of Stephen Hawking. Hertog noted that Hawking "developed his geometric approach to quantum gravity during the period when he was losing the use of his hands for writing equations. This loss may well have encouraged him in his attempt to cast the unfathomable quantum realm of gravity in the language of geometry and topology, which he could visualize on the blackboard."

2. On this, it is also interesting to note that imaginary numbers, introduced around the seventeenth century as cause and effect of the broken link traditionally connecting algebra and geometry, today sit at the core of quantum physics.

3. In fact, as it will become clearer in the following chapters, the body too can be regarded as a technology establishing its own proper ecology (cf. Newen et al. 2018).

4. As all theories of origin, this one, almost legendary, is disputed. See, for instance, Gong Yushu (2010).

5. There is an argument about whether Chinese is really monosyllabic because so many words actually have more than one syllable. For the present analysis, the argument is a moot point. Chinese, by comparison to all the other languages mentioned above, has a very large number of syllables such as *ma* or *shi* that can each mean more than forty different words. Nor does it help to call these "morphemes" precisely for the fact that even if they are customarily combined with other words, they just as frequently appear on their own, hence needing phonological markers to distinguish among their possible meanings.

6. For example, there is no way one could read a Phoenician text without knowing Phoenician, whereas it is possible to read any language written alphabetically without knowing it beforehand.

7. In *Technics and Time: Disorientation*, Stiegler (2009, 68) explains this idea very clearly: "'Proper writing' ('*l'écriture proprement dite*') is what is readable as a result of us having at our disposal the recording 'code.' Pictographic tables remain unreadable for us even when we have the code at our disposal: one must also have knowledge of the context. Without this, the signification escapes."

8. See the lecture by Lera Boroditsky, "How Language Shapes Thought," https://www.youtube.com/watch?v=I64RtGofPW8.

9. In fact, della Chiesa (2010, 142) subsequently calls for a "transdisciplinary and cross-cultural effort" that resonates with what we will be discussing in chapter 5.

Chapter 2

1. Stephen Buranyi, "The WHO v Coronavirus: Why It Can't Handle the Pandemic," *The Guardian*, April 10, 2020, https://www.theguardian.com/news/2020/apr/10/world-health-organization-who-v-coronavirus-why-it-cant-handle-pandemic.

2. Stephen Buranyi, "The WHO v Coronavirus: Why It Can't Handle the Pandemic."

3. A video of the conference can be found on YouTube: https://youtu.be/rydvSn7bBbY.

4. This discussion will focus primarily on Western societies, although comparisons with Asian ones will surface, bringing to light, once again, technocultural differences. In fact, the loss of meaning and the referent, discussed below, is a different issue in logograph-based cultures and this might also be one of the reasons why Asian cultures are still more collective. For instance, Min Wang and colleagues (2020) show that fake news in China is primarily focused on social life–related news (rather than politics) and often in the form of unsourced rumors (rather than fabricated to appear credible). Having said this, however, one of the consequences of

the digital ecology is to reduce language-based technocultural differences, through increasing formatting of /reality/.

5. In the codex book, the printing system standardized the appearance of text with the consequence to disjoin writing from the cognitive-gestural mastering of the human mind and body and to externalize its results as a by-product of a mechanical process. Painting followed a similar path with the introduction of photography: if painters are still the master of their tools, photographers delegate the representation of reality to a mechanical process, which makes everything that appears in a photograph significant, precisely for the very fact of being there regardless of the photographer's intervention.

6. The term, merging "information" and "epidemics" and signaling an overload of often unchecked information, was first coined by journalist David J. Rothkopf in a 2003 article in *The Washington Post*, http://www1.udel.edu/globalagenda/2004 /student/readings/infodemic.html.

7. Umair Haque, "Britain is Self-Destructing—And It's a Warning to the World," Eudaimonia&Co., May 8, 2021, https://eand.co/britain-is-self-destructing-and-its-a -warning-to-the-world-b492870ffc28.

8. Michael Safi, Arun Budhathoki, and India Rakusen, "'Walking over Bodies': Moun-taineers Describe Carnage on Everest," *The Guardian*, May 28, 2019, https://www .theguardian.com/world/2019/may/28/walking-over-bodies-mountaineers-describe -the-carnage-at-the-top-of-mount-everest.

9. For example, in class, he would compare Georges Rouault's paintings as recovering medieval stained-glass practices to Georges Seurat's pointillism to predicting television.

10. Cf. Lee Smolin, "Think About Nature," *The Edge*, https://www.edge.org/conver sation/lee_smolin-think-about-nature

Chapter 3

1. In fact, we do not claim that datacracy is an inescapable scenario. There are indeed various tactics to contrast and rework datafication from the inside (for an overview, cf. Hesselberth 2018). Two main kinds of tactics can be identified: "re-action" and "pro-actions." Re-actions are those interventions that bear an eminently oppositional stance toward digital transformation. These reactions remain framed within a logic of "response" to digital transformation, without really tackling or undermining its raison d'être. Pro-actions, by contrast, can be considered those actions that harness the resources of the digital transformation *in order to* reshape it from the inside. There are various strategies and varying degrees of activism by means of which people can counteract their own commodification as data subjects (cf. Daly et al. 2019).

2. In fact, echoing Appadurai's distinction of technology as both "mechanical" and "informational," it is possible to regard data as *Janus-faced techno-informational con-structs*: if one stresses the *informational* constituency of data, then data are a virtual

entity and are potentially distributable globally; if one stresses the *technical* constituency of data (from collection to storing and use), then data are material entities whose allocation and circulation can be favored or hindered in many ways. This is reflected in different legal understandings and frameworks (e.g., European vs. US doctrines).

3. Charles Duhigg, "How Companies Learn Your Secrets," *The New York Times*, February 16, 2012, https://www.nytimes.com/2012/02/19/magazine/shopping-habits.html.

4. "BRICS" stands for Brazil, Russia, India, China, and South Africa; "LMICs" stands for "low- and middle-income countries."

5. Greg Bensinger, "Google Redraws the Borders on Maps Depending on Who's Looking," *The Washington Post*, February 14, 2020, https://www.washingtonpost.com/technology/2020/02/14/google-maps-political-borders/.

6. According to neuroscience, this is favored by the development of the neocortex, which has plasticity, meaning that it can "adapt" to external inputs, while, for instance, in chimpanzees, the neocortex is much more defined by the genes, making it less malleable (cf. Aida Gómez-Robles et al. 2015).

7. Marshall McLuhan, "A Media Approach to Inflation," *The New York Times*, September 21, 1974. https://www.nytimes.com/1974/09/21/archives/a-media-approach-to-inflation.html

8. Roberto Saracco, "Exploring Ideas to Foster the Metaverse—XVIII," *IEEE Future Directions*, June 1, 2023, https://cmte.ieee.org/futuredirections/2023/06/01/exploring-ideas-to-foster-the-metaverse-xviii/.

9. See European Commission, *Developing CitiVerse*, https://digital-strategy.ec.europa.eu/en/events/info-day-developing-citiverse.

10. Grand View Research, "Smart Cities Market Analysis Report by Application, by Region, and Segment Forecasts 2019–2025," https://www.grandviewresearch.com/industry-analysis/smart-cities-market.

11. See "Symbiotic Autonomous Systems," *IEEE Digital Reality*, November 2019, https://digitalreality.ieee.org/images/files/pdf/1SAS_WP3_Nov2019.pdf.

12. A correlated topic, which we do not have the space to expand upon, is *transhumanism*, that is, "a transition to a new species. It is looking not at the symbioses between us and our artifacts but at the possibility of changing the characteristics (or some of them) of the human race" (Saracco 2018). From the perspective of symbiotic autonomous systems, and also building upon the ideas discussed in chapter 1, *transhumanism* shows a deterministic fallacy, notably a linear understanding of evolution, which essentializes and normalizes "the human" as the supposedly natural benchmark for all physical and cognitive performances, as well as "technology" as a set of plugins to be implemented onto/into the natural human body. As Rosi Braidotti writes in "Posthuman, All Too Human: Towards a New Process Ontology" (2006, 197), "In the historical era of advanced postmodernity, the very notion of 'the human' is not only destabilized

by technologically mediated social relations in a globally connected world, but it is also thrown open to contradictory redefinitions of what exactly counts as human." In fact, we might say, humans have never reached a posthuman condition simply because they have always been transhuman (i.e., technological beings) in the first place.

Chapter 4

1. "Cloud computing" is a tech solution typically good for accumulating and storing huge amounts of data off-site, although their processability may require some time: it is a matter of seconds and even less, but in certain scenarios, this might be crucial. "Edge computing" configures a "lighter" form of data accumulation and also one that is faster to process because data are kept at the edge of the network, yet infrastructuring costs might be higher. "Fog computing" is a horizontal architecture somewhat in between the other two, which, before processing, helps filter data.

2. Zack Witthaker, "NSAs Is So Overwhelmed with Data, It's No Longer Effective, Says Whistleblower," *ZDNet*, April 27, 2016, 2015, https://www.zdnet.com/article /nsa-whistleblower-overwhelmed-with-data-ineffective.

3. Scott Aaronson, "Why Google's Quantum Supremacy Milestone Matters," *The New York Times*, October 30, 2019, https://www.nytimes.com/2019/10/30/opinion /google-quantum-computer-sycamore.html.

4. For a detailed review of the experiment, see https://scottaaronson.blog/?p=4372. One year later, in December 2020, a Chinese research team from the University of Science and Technology of China in Hefei claimed to have built a quantum computer, named Jiuzhang, which is ten billion times faster than Sycamore.

5. Scott Aaronson, "Why Google's Quantum Supremacy Milestone Matters."

6. Scott Aaronson, "The Limits of Quantum Computers," *Scientific American*, March 2008, https://www.cs.virginia.edu/~robins/The_Limits_of_Quantum_Computers.pdf.

7. In 1993, computer scientists defined a class of problems called BQP ("bounded-error quantum polynomial time"), which encompasses all problems that quantum computers can solve. This class is opposed to PH problems ("polynomial hierarchy"), which are those problems solvable by traditional computers (even those still to be invented, but always based on classical physics). Since then, computer scientists have looked for a problem that could be part of BQP and not of PH, meaning that it was solvable *only* by a quantum computer, regardless of the crunching power of a classical one. In 2018, computer scientists Ran Raz from Princeton University and Avishay Tal from Stanford University managed to do that. They built upon the "for-relation" problem elaborated in 2009 by Aaronson and already proven to belong to BQP: this problem asks to what extent two sequences of numbers can be said to be completely independent from each other. What Raz and Tal did was to prove that Aaronson's problem is not a PH problem but is uniquely a BPQ problem. This means that it cannot be solved by traditional computers, only by quantum computers.

8. See https://globalriskinstitute.org/publication/quantum-computing-cybersecurity/.

9. See "Quantum Flagship—The Future Is Quantum," https://qt.eu/.

10. See "Quantum Flagship—The Future Is Quantum," https://qt.eu/.

11. See "Supporting Quantum Technologies Beyond H2020," https://qt.eu//app/uploads/2020/04/Strategic_Research-_Agenda_d_FINAL.pdf.

12. See "Supporting Quantum Technologies Beyond H2020," https://qt.eu//app/uploads/2020/04/StrategicResearch-AgendadFINAL.pdf

13. See "Supporting Quantum Technologies Beyond H2020," https://qt.eu//app/uploads/2020/04/Strategic_Research-_Agenda_d_FINAL.pdf.

14. See "Supporting Quantum Technologies Beyond H2020," https://qt.eu//app/uploads/2020/04/Strategic_Research-_Agenda_d_FINAL.pdf.

15. See "National Strategic Overview for Quantum Information Science," https://www.quantum.gov/wp-content/uploads/2020/10/2018_NSTC_National_Strategic_Overview_QIS.pdf.

16. See "National Strategic Overview for Quantum Information Science," https://www.quantum.gov/wp-content/uploads/2020/10/2018_NSTC_National_Strategic_Overview_QIS.pdf.

17. Jon Lindsay, "Why Is Trump Funding Quantum Computing Research but Cutting Other Science Budgets?" *The Washington Post*, March 12, 2020, https://www.washingtonpost.com/politics/2020/03/13/why-is-trump-funding-quantum-computing-research-cutting-other-science-budgets/.

18. See "A Strategic Vision for America's Quantum Networks," https://www.quantum.gov/wp-content/uploads/2021/01/A-Strategic-Vision-for-Americas-Quantum-Networks-Feb-2020.pdf.

19. See "A Strategic Vision for America's Quantum Networks," https://www.quantum.gov/wp-content/uploads/2021/01/A-Strategic-Vision-for-Americas-Quantum-Networks-Feb-2020.pdf.

20. See "A Strategic Vision for America's Quantum Networks," https://www.quantum.gov/wp-content/uploads/2021/01/A-Strategic-Vision-for-Americas-Quantum-Networks-Feb-2020.pdf.

21. See "13th Five-Year Plan for Economic and Social Development of the People's Republic of China (2016–2021)," https://en.ndrc.gov.cn/policies/202105/P020210527785800103339.pdf.

22. See "13th Five-Year Plan for Economic and Social Development of the People's Republic of China (2016–2021)," https://en.ndrc.gov.cn/policies/202105/P020210527785800103339.pdf.

23. See https://www3.weforum.org/docs/WEF_Quantum_Computing_2022.pdf.

Chapter 5

1. Note that these four possibilities are not gradients: knowledge is not a path from D to A. Rather, they are codependent, with D being an analytical concept that surrounds the other three, C the space of possibility and probability, B the space of aware enaction, and A the space of certainty.

2. See, "BackReAction," https://backreaction.blogspot.com/search?q=eraser+experi ment.

3. See, "GYOC," https://growyourown.cloud/.

4. Note that in the two cases, the interfering patterns are slightly shifted from each other—wave peaks of one pattern correspond to wave valleys in the other pattern—so that, when taken together, the particle-like behavior of the photons is actually restored.

5. Along a similar line, it is worth mentioning the dialogue between physicist Wolfgang Pauli and psychoanalyst Carl Jung. About their epistolary exchange, Harald Atmanspacher and Christopher Fuchs (2017) claim it contains a proper "conjecture" concerning the possible quantum-based theory for the mind–body problem, in the form of a dual-aspect monism whereby reality is fundamentally regarded as a psychophysically neutral whole, with the mental and the physical emerging out of it. The entanglement between mental and physical states can give rise to synchronous acausal correlations, and they do so on a qualitative—rather than statistical—basis.

6. For instance, notions such as "transparency," "openness," and "privacy" can no longer be taken as one-dimensional: any one of them always presupposes its own opposite. Thus, there cannot be openness (e.g., of data) and transparency, without defining, acknowledging, and accounting for closure and opacity. The same goes for personal data: Nadezhda Purtova (2017) rightly claims (also referring to the double nature of light) that "just as light sometimes acts as a particle and sometimes as a wave, data sometimes act as personal data and at other times as non-personal data." Technology and language share the same logic of parceling of experience, but technological advancement remolds /reality/ in idiosyncratic ways, widening the gap with language. At stake is the fundamental awareness that there is no clear-cut way to discern once and for all whether a certain set of data contains personal data or not; these are two complementary features. Speaking of Open Government Data, Bates (2014) notes that "the ends to which openness is being driven by different social actors have become more complex and contested. For some advocates this emerging complexity has been framed in terms of the 'unintended consequences' of OGD." Educationally speaking, it is precisely these "unintended consequences" that need to become the focus of attention: they are not happening "by chance"; they are systemic.

7. Most important, the "why" of such a transition is worth stressing. It is not only an ethical matter to be at stake; ecological thinking and practice also have socioeconomic benefits. In fact, it has been shown that a shift toward sustainable models could lead to up to $26 trillion savings by 2030 for the whole world (see https:// newclimateeconomy.report/2018/).

8. See https://wiki.p2pfoundation.net/Importance_of_Neotraditional_Approaches_in _the_Reconstructive_Transmodern_Era.

9. For a discussion on the impact of AI on the job market worldwide and on the benefits and pitfalls of possible countermeasures such as the Universal Basic Income, cf. Lee (2018).

10. See "Unleashing the Full Power of Quantum Technologies," https://www.tudelft .nl/over-tu-delft/strategie/vision-teams/quantum-internet/basics-of-quantum -mechanics/future-scenarios.

11. From the Greek *tropein*, to move around something, orient oneself toward an objective, like the flower toward the sun.

12. See https://libbyheaney.co.uk/artworks/ent-many-paths-version/.

References

Aaronson, Scott. 2013. *Quantum Computing Since Democritus*. Cambridge: Cambridge University Press.

Abbott, Derek, Paul Davies, and Arun Pati, eds. 2008. *Quantum Aspects of Life*, London: Imperial College Press.

Aboitiz, Francisco. 2012. "Gestures, Vocalizations, and Memory in Language Origins." *Frontiers in Evolutionary Neuroscience* 4: 2. https://doi.org/10.3389/fnevo.2012.00002.

Accoto, Cosimo. 2017. *In Data Time and Tide*. Milano: Bocconi University Press.

Adorno, Theodor. 1991. *The Culture Industry: Selected Essays on Mass Culture*. London: Routledge.

Adorno, Theodor, and Walter Benjamin. 1999. *The Complete Correspondence: 1928–1940*. London: Polity Press.

Aerts, Diederik, Andrei Khrennikov, Massimo Melucci, and Bourama Toni, eds. 2019. *Quantum-like Models for Information Retrieval and Decision-making*. Cham: Springer Nature.

Aerts, Diederik, Marek Czachor, and Bart D'Hooghe. 2006. "Towards a Quantum Evolutionary Scheme: Violating Bell's Inequalities in Language." In *Evolutionary Epistemology, Language and Culture: A Non Adaptationist Systems Theoretical Approach*, edited by Nathalie Gontier, Jean Paul van Bendegem, and Diederik Aerts, 453–478. Heidelberg: Springer.

Agamben, Giorgio. 2009. *What Is an Apparatus?* Translated by David Kishik and Stefan Pedatella. Stanford: Stanford University Press.

Ahrens, Kathleen, and Chu-Ren Huang. 2002. "Time Is Motion in Mandarin Chinese. Parameterizing Conceptual Metaphors." In *Proceedings of the 7th International Symposium of Chinese Languages and Linguistics*, 22–47. Jiayi: National Chung-Cheng University.

Al-Khalili, Jim, and Johnjoe McFadden. 2015. *Life on the Edge: The Coming of Age of Quantum Biology*. London: Bantam Press.

Albert, Mathias, and Felix M. Bathon. 2020. "Quantum and Systems Theory in World Society: Not Brothers and Sisters but Relatives Still?" *Security Dialogue* 51 (5): 434–449.

Alexander, Bruce K., Barry L. Beyerstein, Patricia F. Hadaway, and Robert B. Coambs. 1981. "Effect of Early and Later Colony Housing on Oral Ingestion of Morphine in Rats." *Pharmacology Biochemistry and Behavior* 15: 571–576.

Alverson, Hoyt. 1994. *Semantics and Experience: Universal Metaphors of Time in English, Mandarin, Hindi, and Sesotho*. Baltimore: Johns Hopkins University Press.

Amoore, Louise. 2019. "Doubt and the Algorithm: On the Partial Accounts of Machine Learning." *Theory, Culture & Society* 36 (6): 147–169.

Anderson, Chris. 2008. "The End of Theory: The Data Deluge Makes the Scientific Method Obsolete." *Wired Magazine*. http://statlit.org/pdf/2008EndOfTheory-DataDe lugeMakesScientificMethodObsolete-WiredMagazine.pdf.

Appadurai, Arjun. 1990. "Disjuncture and Difference in the Global Cultural Economy." *Theory, Culture & Society* 7: 295–310.

Arnstein, Sherry R. 1969. "A Ladder of Citizen Participation." *Journal of the American Institute of Planners* 35 (4): 216–224.

Atkinson, Quentin. 2011. "Phonemic Diversity Supports a Serial Founder Effect Model of Language Expansion from Africa." *Science Magazine* 332 (6027): 346–349.

Atmanspacher, Harald. 2017. "Quantum Approaches to Brain and Mind." In *The Blackwell Companion to Consciousness*, edited by Susan Schneider and Max Velmans, 298–313. Hoboken, NJ: John Wiley & Sons.

Atmanspacher, Harald. 2020. "The Pauli–Jung Conjecture and Its Relatives: A Formally Augmented Outline." *Open Philosophy* 3 (1): 527–549.

Atmanspacher, Harald, and Christopher A. Fuchs, eds. 2017. *The Pauli-Jung Conjecture and Its Impact Today*. Exeter: Imprint Academic.

Austin, Reginald Percy. 1938. *The Stoichedon Style in Greek Inscriptions*. Oxford: Oxford University Press.

Bailey, Rowan, Claire Booth-Kurpnieks, Kath Davies, and Ioanni Delsante. 2019. "Cultural Ecology and Cultural Critique." *Arts* 8 (166): 1–19.

Bakhtin, Mikhail. 1981. *The Dialogic Imagination: Four Essays*. Translated by Michael Holquist and Caryl Emerson. Austin: University of Texas Press.

Ball, Philip. 2011. "The Dawn of Quantum Biology." *Nature* 474: 272–274.

Barad, Karen, 2003. "Posthumanist Performativity: Toward an Understanding of How Matter Comes to Matter." *Journal of Women in Culture and Society* 28 (3): 801–831.

Barad, Karen. 2007. *Meeting the Universe Halfway, Quantum Physics and the Entanglement of Matter and Meaning*. Durham, NC: Duke University Press.

Barrett, Jeffrey. 2014. "On the Coevolution of Theory and Language and the Nature of Successful Inquiry." *Erkenntnis* 79 (Suppl. 4): 821–834.

Barthes, Roland. 1953. *Le degré zéro de l'écriture*. Paris: Le seuil.

Barthes, Roland. 1972. *Critical Essays*. Translated by Richard Howard. Louisville, KY: Evanston.

Basualdo, Pedro Alejandro. 2011. "Individual Global Responsibility." *United Nations Chronicle*. https://www.un.org/en/chronicle/article/individual-global-responsibility.

Bates, Jo. 2014. "Open Government Data and the Neoliberal State." https://blogs .lse.ac.uk/impactofsocialsciences/2014/10/02/open-government-data-and-the -neoliberal-state/.

Bates, Jo. 2018. "The Politics of Data Friction." *Journal of Documentation* 74 (2): 412–429.

Bateson, Gregory. 2000. *Steps to an Ecology of Mind: Collected Essays in Anthropology, Psychiatry, Evolution, and Epistemology*. Chicago: University of Chicago Press.

Basieva, Irina, Andrei Khrennikov, and Masanao Ozawa. 2021. "Quantum-like Modeling in Biology with Open Quantum Systems and Instruments." *Biosystems* 201: 104328. https://doi.org/10.1016/j.biosystems.2020.104328.

Baudrillard, Jean. 1994. *Simulacra and Simulation*. Ann Arbor: University of Michigan Press.

Bauman, Zygmunt. 1989. *Modernity and the Holocaust*. New York: Cornell University Press.

Bauwens, Michel, Vasilis Kostakis, and Alex Pazaitis. 2019. *Peer to Peer: The Commons Manifesto*. London: University of Westminster Press.

Beer, Kerstin, et al. 2020. "Training Deep Quantum Neural Networks." *Nature Communication* 11 (808). https://doi.org/10.1038/s41467-020-14454-2.

Benjamin, Walter. 1999. "Surrealism: The Last Snapshot of the European Intelligentsia." In *Walter Benjamin: Selected Writings*, edited by Marcus Bullock and Michael W. Jennings, vol. 2.1, 207–221. Cambridge, MA: Belknap Press.

Benjamin, Walter. 2002. "The Storyteller: Observations on the Works of Nikolai Leskov." In *Walter Benjamin: Selected Writings*, edited by Marcus Bullock and Michael W. Jennings, vol. 3, 143–166. Cambridge, MA: Belknap Press.

Berardi, Franco. 2015. *And: Phenomenology of the End*. Cambridge, MA: MIT Press.

Bergson, Henri. 1965. *Duration and Simultaneity with Reference to Einstein's Theory*. Translated by Leon Jacobson. Indianapolis, IN: Bobbs-Merrill.

Bettencourt, Luis M. 2015. "Cities as Complex Systems." In *Modelling Complex Systems for Public Policies*, edited by Bernardo Alves Furtado, Patrícia A. M. E. Sakowski, and Marina H. E. Tóvolli, 217–236. Brasilia: Ipea.

Bibri, Simon Elia. 2018. "The IoT for Smart Sustainable Cities of the Future: An Analytical Framework for Sensor-Based Big Data Applications for Environmental Sustainability." *Sustainable Cities and Society* 38: 230–258.

Bitbol, Michel. 2008. "Is Consciousness Primary?" *NeuroQuantology*, 6 (1), 53–71.

Blanchot, Maurice. 1993. *The Infinite Conversation*. Translated by Susan Hanson. Minneapolis: University of Minnesota Press.

Blatt, Rainer. 2020. "Advancing Quantum Technologies—Chances and Challenges." *Advanced Quantum Technologies* 3: 1.

Bluvstein, Dolev, et al. 2023. "Logical Quantum Processor Based on Reconfigurable Atom Arrays." *Nature* 626: 58–65. https://doi.org/10.1038/s41586-023-06927-3.

Bohm, David. 1952. "A Suggested Interpretation of the Quantum Theory in Terms of 'Hidden' Variables, Parts I and II." *Physical Review* 85: 166–193.

Bohm, David. (1980) 2002. *Wholeness and the Implicate Order*. New York: Routledge Classics.

Bohr, Niels. 1935. "Can Quantum-Mechanical Description of Physical Reality Be Considered Complete?" *Physical Review* 48 (8): 696–702.

Bohr, Niels. 1949. "Discussion with Einstein on Epistemological Problems in Atomic Physics." In *Albert Einstein: Philosopher-Scientist*, edited by Paul Schilpp. New York: MJF Books.

Bohr, Niels. 1987. *The Philosophical Writings of Niels Bohr*. Woodbridge: Ox Bow Press.

Bollier, David. 1998. *How Smart Growth Can Stop Sprawl: A Fledgling Citizen Movement Expands*. Washington, D.C.: Essential Books.

Bollier, David, and Silke Helfrich. 2020. *Free, Fair, and Alive: The Insurgent Power of the Commons*. Gabriola Island: New Society Publishers.

Bookheimer, Susan. 2002. "Functional MRI of Language: New Approaches to Understanding the Cortical Organization of Semantic Processing." *Annual Review of Neuroscience* 25: 151–188.

Bortoft, Henry. 1985. "Counterfeit and Authentic Wholes." In *Dwelling, Place and Environment*, edited by David Seamon and Robert Mugerauer, 281–302. Dordrecht: Martinus Nijhoff.

Boschert, Stefan, and Roland Rosen. 2016. "Digital Twin: The Simulation Aspect." In *Mechatronic Futures: Challenges and Solutions for Mechatronic Systems and Their Designers*, edited by Peter Hehenberger and David Bradley, 59–74. Berlin: Springer.

Bradley, Arthur. 2011. *Originary Technicity: The Theory of Technology from Marx to Derrida*. Basingstoke: Palgrave/MacMillan.

Braidotti, Rosi. 2006. "Posthuman, All Too Human towards a New Process Ontology." *Theory, Culture & Society* 23 (7–8): 197–208.

Brannon, Monica. 2017. "Datafied and Divided: Techno-Dimensions of Inequality in American Cities." *City & Community* 16 (1): 20–24.

Bratton, Benjamin. 2016. *The Stack.* Cambridge, MA: MIT Press.

Burwell, Jennifer. 2018. *Quantum Language and the Migration of Scientific Concepts.* Cambridge, MA: MIT Press.

Busemeyer, Jerome, and Peter Bruza. 2012. *Quantum Models of Cognition and Decision.* Cambridge: Cambridge University Press.

Butor, Michel. 1957. *La modification.* Paris: Nathan.

Calzati, Stefano. 2018. "A Proposal for Survival: Barbaric Strategies in the Realm of Digital Technologies." *Parallax* 24 (2): 209–226.

Calzati, Stefano. 2020. "Decolonising 'Data Colonialism': Propositions for Investigating the Realpolitik of Today's Networked Ecology." *Television & New Media.* https://doi.org/10.1177/1527476420957267.

Calzati, Stefano. 2023. "Shaping a Data Commoning Polity: Prospects and Challenges of a European Digital Sovereignty." In *International Conference on Electronic Participation*, 151–166. Cham: Springer.

Calzati, Stefano, and Bastiaan van Loenen. 2023a. "A Fourth Way to the Digital Transformation: The Data Republic as a Fair Data Ecosystem." *Data & Policy* 5: e21.

Calzati, Stefano, and Bastiaan van Loenen. 2023b. "Beyond Federated Data: A Data Commoning Proposition for the EU's Citizen-centric Digital Strategy." *AI & Society.* https://doi.org/10.1007/s00146-023-01743-9.

Calzati, Stefano, and Bastiaan van Loenen. 2023c. "Towards a Citizen- and Citizenry-Centric Digitalization of the Urban Environment: Urban Digital Twinning as Commoning." *Digital Society.* https://doi.org/10.1007/s44206-023-00064-0.

Calzati, Stefano, and Roberto Simanowski. 2018. "Self-Narratives on Social Networks: Trans-Platforms Stories and Facebook's Metamorphosis into a Postmodern Semi-Automated Repository." *Biography: An Interdisciplinary Quarterly* 41 (1): 24–47.

Campagne-Ibarcq, Phillipe, et al. 2020. "Quantum Error Correction of a Qubit Encoded in Grid States of an Oscillator." *Nature* 584: 368–372.

Cao, Cong, Jeroen Baas, Caroline Wagner, and Koen Jonkers. 2020. "Returning Scientists and the Emergence of China's Science System." *Science and Public Policy* 47 (2): 172–183.

Capra, Fritjof. 1975. *The Tao of Physics.* Boulder, CO: Shambhala Publications.

Cardullo, Paolo, and Rob Kitchin. 2019. "Smart Urbanism and Smart Citizenship: The Neoliberal Logic of 'Citizen-Focused' Smart Cities in Europe." *Environment and Planning C: Politics and Space* 37 (5): 813–830.

Castelli, Alberto. 2015. "On Western and Chinese Conception of Time: A Comparative Study." *Philosophical Papers and Reviews* 6 (4): 23–30.

Castells, Manuel. 1998. *End of Millennium: The Information Age: Economy, Society, and Culture*. Vol. 3. Oxford: Blackwell.

Castells, Manuel. 2007. "Communication, Power and Counter-power in the Network Society." *International Journal of Communication* 1: 238–266.

Changeux, Jean-Pierre. 1997. *Neuronal Man: The Biology of Mind*. Princeton, NJ: Princeton University Press.

Chomsky, Noam. 1996. *Powers and Prospects. Reflections on Human Nature and the Social Order*. London: Pluto Press.

Coenen, Christopher, and Armin Grunwald. 2017. "Responsible Research and Innovation in Quantum Technology." *Ethics and Information Technology* 19 (4): 277–294.

Cohen, Peter. 2009. "The Naked Empress. Modern Neuro-science and the Concept of Addiction." In *12th Platform for Drug Treatment*, Mondsee Austria, 21–22 March. http://www.antoniocasella.eu/dnlaw/Cohen_2009.pdf.

Da-Wei, Kwo. 1990. *Chinese Brushwork in Calligraphy and Painting*. New York: Dover.

Daly, Angela, Kate Devitt, and Monique Mann, eds. 2019. *Good Data*. Amsterdam: Institute of Network Cultures.

de Angelis, Massimo. 2017. *Omnia Sunt Communia: On the Commons and the Transformation to Postcapitalism*. London: Zed Books.

de Jong, Eline. 2022. "Own the Unknown: An Anticipatory Approach to Prepare Society for the Quantum Age." *Digital Society* 1, 15. https://link.springer.com/article/10.1007/s44206-022-00020-4.

de Kerckhove, Derrick. 1997a. *The Skin of Culture: Investigating the New Electronic Reality*. London: Kogan Page.

de Kerckhove, Derrick. 1997b. *Connected Intelligence: The Arrival of the Web Society*. Toronto: Somerville.

de Kerckhove, Derrick. 2020. *The Skin of Culture*. Updated Chinese edition, translated by Daokuan He. Beijing: The Chinese Encyclopedia Publishers.

de Kerckhove, Derrick, and Charles Lumsden, eds. 1988. *The Alphabet and the Brain: The Lateralization of Writing*. Berlin: Springer.

de Kerckhove, Derrick, and Cristina Miranda de Almeida, eds. 2014. *The Point of Being*. Newcastle: Cambridge Scholars.

de Saussure, Ferdinand. (1916) 1959. *Course in General Linguistics.* New York: The Philosophical Society.

Debord, Guy. 1967. *La société du spectacle.* Paris: Buchet Chastel.

Dehaene, Stanislas. 2009. *Reading in the Brain: The New Science of How We Read.* New York: Penguin.

Dehaene, Stanislas. 2011. *The Number Sense: How the Mind Creates Mathematics.* New York: Oxford University Press.

Deleuze, Gilles. 1992. "Postscript on the Societies of Control." *October* 59 (Winter): 3–7.

Deleuze, Gilles. 1995. *Negotiations.* New York: Columbia University Press.

Deleuze, Gilles, and Felix Guattari. 1987. *A Thousand Plateaus: Capitalism and Schizophrenia.* Translated by Brian Massumi. London: Continuum.

Delgado, Pablo, Cristina Vargas, Rakefet Ackerman, and Ladislao Salmerón. 2018. "Don't Throw Away Your Printed Books: A Meta-Analysis on the Effects of Reading Media on Reading Comprehension." *Educational Research Review* 25: 23–38.

della Chiesa, Bruno. 2010. "Tesseract: One Hypothesis on Languages, Cultures, and Ethics." *Mind, Brain, and Education* 4 (3): 135–148.

Dembski, Fabian, et al. 2020. "Urban Digital Twins for Smart Cities and Citizens: The Case Study of Herrenberg, Germany." *Sustainability* 12 (6): 2307.

Dencik, Lina, Arne Hintz, and Jonathan Cable. 2016. "Towards Data Justice? The Ambiguity of Anti Surveillance Resistance in Political Activism." *Big Data & Society* 3 (2): 1–12.

Der Derian, James, and Alexander Wendt. 2020. "'Quantizing International Relations': The Case for Quantum Approaches to International Theory and Security Practice." *Security Dialogue* 51 (5): 399–413.

Derrida, Jacques. 1976. *Of Grammatology.* Translated by Gayatri Chakravorty Spivak. Baltimore: Johns Hopkins University Press.

Derrida, Jacques, and Bernard Stiegler. 2002. *Echographies of Television. Filmed Interviews.* Translated by Jennifer Bajorek. Cambridge: Polity Press.

Deutscher, Guy. 2010. *Through the Language Glass: Why the World Looks Different in Other Languages.* London: Picador.

Devereaux, Abigail, Roger Koppl, Stuart Kauffman, and Andrea Roli. 2021. "An Incompleteness Result Regarding Within-System Modeling." *SSRN.* https://dx.doi.org/10.2139/ssrn.3968077.

Devitt, Michael, and Kim Sterelny. 1999. *Language and Reality: An Introduction to the Philosophy of Language.* 2nd ed. Cambridge, MA: MIT Press.

Dirac, Paul. 1929. "Quantum Mechanics of Many-Electron Systems." *Proceedings of the Royal Society A* 123: 714–733.

Dowling, Jonathan, and Gerard Milburn. 2003 "Quantum Technology: The Second Quantum Revolution." *Philosophical Transactions of the Royal Society of London. Series A: Mathematical, Physical and Engineering Sciences* 361: 1655–1674.

Drechsler, Wolfgang. 2019. "Kings and Indicators: Options for Governing Without Numbers." In *Science, Numbers and Politics*, edited by Markus J. Prutsch, 227–262. Basingstoke: Palgrave Macmillan.

Dutton, William, Kenneth Kraemer, and Jay Blumler. 1987. *Wired Cities: Shaping the Future of Communications*. London: Macmillan.

Easterling, Keller. 2021. *Medium Design. Knowing How to Work on the World*. London: Verso.

Eastman, Timothy. 2020. *Untying the Gordian Knot: Process, Reality, and Context*. Lanham, MD: Lexington Books.

Eco, Umberto. 1962. *Opera aperta*. Milano: Bompiani.

Edwards, Paul. 2010. *A Vast Machine: Computer Models, Climate Data, and the Politics of Global Warming*. Cambridge, MA: MIT Press.

Einstein, Albert, Boris Podolsky, and Nathan Rosen. 1935. "Can Quantum-Mechanical Description of Physical Reality Be Considered Complete?" *Physical Review* 47: 777–780.

Eisenstein, Elizabeth. 1979. *The Printing Press as an Agent of Change*. Cambridge: Cambridge University Press.

Elitzur, Avshalom C., and Lev Vaidman. 1993. "Quantum Mechanical Interaction-Free Measurements." *Foundations of Physics* 23: 987–997.

Epstein, Jean. (1946) 2014. *The Intelligence of a Machine*. Translated by Christophe Wall-Romana. Minneapolis, MN: Univocal Publishing.

Esposito, Elena. 2017. "An Ecology of Differences: Communication, the Web, and the Question of Borders." In *General Ecology: The New Ecological Paradigm*, edited by Erich Hörl with James Burton, 285–302. London: Bloomsbury.

Esposito, Roberto. 2022. *Communitas. In Communitas*. Stanford, CA: Stanford University Press.

Eubanks, Virginia. 2018. *Automating Inequality: How High-Tech Tools Profile, Police, and Punish the Poor*. New York: St. Martin's.

European Commission. 2019a. "Ethics Guidelines for Trustworthy AI." https://digital-strategy.ec.europa.eu/en/library/ethics-guidelines-trustworthy-ai.

European Commission. 2019b. "Policy and Investment Recommendations for Trustworthy AI." https://digital-strategy.ec.europa.eu/en/library/policy-and-investment-recommendations-trustworthy-artificial-intelligence.

European Commission. 2021. "Digital Europe Programme (DIGITAL)." https://digital-strategy.ec.europa.eu/en/news/first-calls-proposals-under-digital-europe-programme-are-launched-digital-tech-and-european-digital.

European Commission. 2022. "European Declaration on Digital Rights and Principles for the Digital Decade." https://digital-strategy.ec.europa.eu/en/library/declaration-european-digital-rights-and-principles.

Evans, Nicholas, and Stephen Levinson. 2009. "The Myth of Language Universals: Language Diversity and Its Importance for Cognitive Science." *Behavioral and Brain Sciences* 32: 429–492.

Everett, Hugh. 1957. "Relative State Formulation in Quantum Mechanics." *Reviews of Modern Physics* 29: 454–462.

Fabbri, Paolo. 2001. *La svolta semiotica*. Bari: Laterza.

Festinger, Leon. 1957. *A Theory of Cognitive Dissonance*. Stanford, CA: Stanford University Press.

Feynman, Richard. 1982. "Simulating Physics with Computers." *International Journal of Theoretical Physics* 21 (6–7): 467–488.

Feynman, Richard. 1994. *No Ordinary Genius: The Illustrated Richard Feynman*. Edited by Christopher Sykes. New York: W. W. Norton.

Filipović, Luna, and Kasia Jaszczolt. 2012. *Space and Time in Languages and Cultures*. Amsterdam: John Benjamins.

Floridi, Luciano. 2017. "Digital's Cleaving Power and Its Consequences." *Philosophy of Technology* 30: 123–130.

Floridi, Luciano. 2019. "What the Near Future of Artificial Intelligence Could Be." *Philosophy & Technology* 32: 1–15.

Floridi, Luciano. 2020. *Il verde e il blu: Idee ingenue per migliorare la politica*. Milano: Raffaello Cortina Editore.

Foucault, Michel. 1980. *Power/Knowledge: Selected Interviews and Other Writings, 1972–1977*. Edited by Colin Gordon. New York: Pantheon Books.

Foucault, Michel. 1988. "Technologies of the Self." In *Technologies of the Self: A Seminar with Michel Foucault*, edited by Luther H. Martin, Huck Gutman, Patrick H. Hutton, 16–49. London: Tavistock Publications.

Foucault, Michel. 1997. *Ethics: Subjectivity and Truth*. New York: The New Press.

Frické, Martin. 2009. "The Knowledge Pyramid: A Critique of the DIKW Hierarchy." *Journal of Information Science* 35 (2): 131–142.

Frischmann, Brett M., Michael J. Madison, and Katherine Jo Strandburg, eds. 2014. *Governing Knowledge Commons*. Oxford: Oxford University Press.

Frosh, Paul. 2013. "Beyond the Image Bank: Digital Commercial Photography." In *The Photographic Image in Digital Culture*, edited by Martin Lister, 131–148. London: Routledge.

Fu, Xiaolan, Wing Thye Woo, and Jun Hou. 2016. "Technological Innovation Policy in China: The Lessons, and the Necessary Changes Ahead." *Economic Change and Restructuring* 49: 139–157.

Fuller, Aidan, Zhong Fan, Charles Day, and Chris Barlow. 2020. "Digital Twin: Enabling Technologies, Challenges and Open Research." *IEEE Access* 8: 108952–108971. https://doi.org/10.1109/ACCESS.2020.2998358.

Gabora, Liane, Eleanor Rosch, and Diederik Aerts. 2008. "Towards an Ecological Theory of Concepts." *Ecological Psychology* 20: 84–116.

Gagliardone, Iginio. 2019. *China, Africa, and the Future of the Internet*. London: Zed Books.

Gao, Shan, ed. 2018. *Collapse of the Wave Function: Models, Ontology, Origin, and Implications*. Cambridge: Cambridge University Press.

George, Gerard, Ryan Merrill, and Simon Schillebeeckx. 2020. "Digital Sustainability and Entrepreneurship: How Digital Innovations Are Helping Tackle Climate Change and Sustainable Development." *Entrepreneurship Theory and Practice* 45 (5): 999–1027.

Georgescu, Iulia. 2020. "25 Years of Quantum Error Correction." *Nature Reviews Physics* 2: 519.

Gershenson, Carlos. 2015. "Requisite Variety, Autopoiesis, and Self-Organization." *Kybernetes* 44 (6/7): 866–873.

Ghirardi, Giancarlo, Alberto Rimini, and Thomas Weber. 1986. "Unified Dynamics for Microscopic and Macroscopic Systems." *Physical Review D* 34: 470–490.

Gibson, James J. 1977. "The Theory of Affordances." *Hilldale, USA* 1 (2): 67–82.

Ginsberg, Jeremy, et al. 2009. "Detecting Influenza Epidemics Using Search Engine Query Data." *Nature* 457: 1012–1014.

Giovannini, Enrico. 2016. *L'utopia sostenibile*. Bari: Laterza.

Glasze, Georg, et al. 2023. "Contested Spatialities of Digital Sovereignty." *Geopolitics* 28 (2): 919–958.

Gómez-Robles, Aida, William D. Hopkins, Steven J. Schapiro, and Chet C. Sherwood. 2015. "Relaxed Genetic Control of Cortical Organization in Human Brains Compared with Chimpanzees." *PNAS* 112 (48): 14799–14804.

Gong, Ping. 2009. "Study of the Processing Patterns of Time as Space Metaphors." *Journal of Yangtze University* 32: 97–99.

Google Quantum AI. 2023. "Suppressing Quantum Errors by Scaling a Surface Code Logical Qubit." *Nature* 614: 676–681.

Gottesman, Daniel, Alexei Kitaev, and John Preskill. 2001. "Encoding a Qubit in an Oscillator." *Physical Review A* 64 (1): 012310.

Graziano, Michael. 2013. *Consciousness and the Social Brain*. Oxford: Oxford University Press.

Greenstein, George, and Arthur G. Zajonc. 1997. *The Quantum Challenge: Modern Research on the Foundations of Quantum Mechanics*. Sudbury: Jones and Bartlett.

Grieves, Michael. (2002) 2014. "Digital Twin: Manufacturing Excellence through Virtual Factory Replication." *White Paper* 1 (2014): 1–7.

Grieves, Michael, and John Vickers. 2017. "Digital Twin: Mitigating Unpredictable, Undesirable Emergent Behaviour in Complex Systems." In *Transdisciplinary Perspectives on Complex Systems*, edited by Franz-Josef Kahlen, Shannon Flumerfelt, and Anabela Alves, 85–113. Cham: Springer.

Grimstrup, Jesper. 2021. *Shell Beach: The Search for The Final Theory*. Crowdfunded. Kindle.

Grinbaum, Alexei. 2017. "Narratives of Quantum Theory in the Age of Quantum Technologies." *Ethics and Information Technology* 19 (4): 295–306.

Guarducci, Marguerita. 2005. *L'epigrafia greca dalle origini al tardo impero*. Roma: Edizioni Istituto Poligrafico dello Stato.

Guattari, Félix. 1995. *Chaosmosis: An Ethico-Aesthetic Paradigm*. Translated by Paul Bains and Julian Pefanis. Bloomington: Indiana University Press.

Gumbrecht, Hans-Ulrich. 2012. *Atmosphere, Mood, Stimmung on a Hidden Potential of Literature*. Stanford, CA: Stanford University Press.

Habermas, Jurgen. 1989. *The Structural Transformation of the Public Sphere: An Inquiry into a Category of Bourgeois Society*. Cambridge, MA: MIT Press.

Hadamard, Jacques. 1949. *An Essay on the Psychology of Invention in the Mathematical Field*. Princeton, NJ: Princeton University Press.

Han, Byung-Chul. 2016. *Le Parfum du temps*. Paris: Edition Circé.

Han, Sufen, Hongyu Liu, and Yan Lin. 2019. "Measurement of the Innovation Efficiency of the Hi-tech Industry in China and Its Influencing Factors." *International Journal of Sustainable Development and Planning* 15 (3): 277–286.

Harari, Yuval Noah. 2016. *Homo Deus: A Brief History of Tomorrow*. London: Harvill Secker.

Haraway, Donna. 1988. "Situated Knowledges: The Science Question in Feminism and the Privilege of Partial Perspective." *Feminist Studies* 14 (3): 575–99.

Hasselbalch, Gry. 2020. "Culture by Design: A Data Interest Analysis of the European AI Policy Agenda." *First Monday.* https://doi.org/10.5210/fm.v25i12.10861.

Hasselbalch, Gry. (2022). "Data Pollution & Power–White Paper." https://uni-bonn.sciebo.de/s/bYOsFyNiZ9sPlY4/download.

Havelock, Eric. 1976. *Origins of Western Literacy.* Ontario: Ontario Institute for Studies in Education.

Havelock, Eric. 1982. *Preface to Plato.* Cambridge, MA: Harvard University Press.

Haven, Emmanuel, and Andrei Khrennikov. 2013. *Quantum Social Science.* Cambridge: Cambridge University Press.

Hayles, Katherine. 2012. *How We Think Digital Media and Contemporary Technogenesis.* Chicago: University of Chicago Press.

Heidegger, Martin. 1977. *The Question Concerning Technology.* Translated by William Lovitt. New York: Garland Publishing.

Heidegger, Martin. 2004. *The Phenomenology of Religious Life.* Translated by Matthias Fritsch and Jennifer Anna Gosetti-Ferencei. Bloomington: Indiana University Press.

Heisenberg, Werner. 1925. "On a Quantum-Theoretical Reinterpretation of Kinematics and Mechanical Relations." Reprinted and translated in Bartel Leendert van der Waerden (1967) *Sources of Quantum Mechanics.* New York: Dover.

Heisenberg, Werner. 1971. *Physics and Beyond: Encounters and Conversations.* London: Allen and Unwin.

Hertog, Thomas. 2023. *On the Origin of Time: Stephen Hawking's Final Theory.* New York: Bantam.

Hess, Charlotte, and Elinor Ostrom. 2007. *A Framework for Analyzing the Knowledge Commons.* Cambridge, MA: MIT Press.

Hesselberth, Pepita. 2018. "Discourses on Disconnectivity and the Right to Disconnect." *New Media & Society* 20 (5): 1994–2010.

Hidalgo, César. 2015. *Why Information Grows: The Evolution of Order, from Atoms to Economies.* New York: Basic Books.

Hilty, Lorentz, Claudia Som, and Andreas Koehler. 2004. "Assessing the Human, Social, and Environmental Risks of Pervasive Computing." *Human and Ecological Risk Assessment: An International Journal* 10: 853–874.

Hintz, Arne, Lina Dencik, and Karin Wahl-Jorgensen. 2017. "Digital Citizenship and Surveillance Society: Introduction." *International Journal of Communication* 11: 731–739.

Hofstadter, Douglas. 1979. *Gödel, Escher, Bach: An Eternal Golden Braid.* New York: Basic Books.

Hofstede, Geert. 1980. *Culture's Consequences: International Differences in Work-Related Values*. Beverly Hills, CA: Sage.

Hong, Yu, and G. Thomas Goodnight. 2022. "How to Think about Cyber-sovereignty: The Case of China." In *China's Globalizing Internet: History, Power, and Governance*, edited by Yu Hong and Eric Harwit, 7–25. London: Routledge.

Hoofnagle, Chris Jay, and Simson L. Garfinkel. 2022. *Law and Policy for the Quantum Age*. Cambridge: Cambridge University Press.

Hörl, Erich. 2017. "Introduction to General Ecology: The Ecologization of Thinking." In *General Ecology: The New Ecological Paradigm*, edited by Erich Hörl with James Burton, 1–74. London: Bloomsbury.

Hsü, Kenneth J. 1999. "Why Isaac Newton Was Not a Chinese." *Online Journal for E&P Geoscientists*. http://www.searchanddiscovery.com/documents/Hsu/newton.htm.

Huang, Yasheng. 2018. "China's Use of Big Data Might Actually Make It Less Big Brother-ish." *MIT Technology Review*, August. https://www.technologyreview.com/2018/08/22/140655/chinas-use-of-big-data-might-actually-make-it-less-big-brother-ish/.

Hwangbo, Alfred B. 1999. "A New Millennium and Feng Shui." *The Journal of Architecture* 4 (2): 191–198.

Iaconesi, Salvatore, and Oriana Persico. 2015. "Il terzo infoscape. Dati, informazioni e saperi nella città e nuovi paradigmi di interazione urbana." In *I media digitali e l'interazione uomo-macchina*, edited by Simone Arcagni, 139–168. Roma: Aracne Editore.

Iaconesi, Salvatore, and Oriana Persico. 2017. "Algorithmic Autobiography. A New Literary Genre." *Digimag* 75 (11): 85–96.

IBM Newsroom. 2023. "IBM Debuts Next-Generation Quantum Processor & IBM Quantum System Two, Extends Roadmap to Advance Era of Quantum Utility." https://newsroom.ibm.com/2023-12-04-IBM-Debuts-Next-Generation-Quantum-Processor-IBM-Quantum-System-Two,-Extends-Roadmap-to-Advance-Era-of-Quantum-Utility.

Ihde, Don, and Lambros Malafouris. 2019. "Homo Faber Revisited: Postphenomenology and Material Engagement Theory." *Philosophy & Technology* 32: 195–214.

Ikeda, Kazuki, and Shoto Aoki. 2022. "Theory of Quantum Games and Quantum Economic Behavior." *Quantum Information Processing* 21 (27). https://doi.org/10.1007/s11128-021-03378-5.

Illiano, Jessica, Marcello Caleffi, Antonio Manzalini, and Angela Sara Cacciapuoti. 2022. "Quantum Internet Protocol Stack: A Comprehensive Survey." *Computer Networks* 213. https://doi.org/10.1016/j.comnet.2022.109092

Innis, Harold. 1972. *Empire and Communication*. Toronto: University of Toronto Press.

Jarvis, Scott, and Aneta Pavlenko. 2007. *Crosslinguistic Influence in Language and Cognition*. New York: Routledge.

Jenkins, Henry, and Mark Deuze. 2008. "Convergence Culture." *Convergence* 14 (1): 5–12.

Jia, Yuxin, and Xuerui Jia. 2005. "Chinese Characters, Chinese Culture and Chinese Mind." *Intercultural Communication Studies* 14 (1): 151–157.

Jiang, Min, and King-Wa Fu. 2018. "Chinese Social Media and Big Data: Big Data, Big Brother, Big Profit?" *Policy & Internet* 10 (4): 372–392.

Jiehong, Li. 2005. "Language, Culture, and Science-Technology Development." *Intercultural Communication Studies* 4 (2): 102–113.

Jindal, Anish, Neeraj Kamur, and Mukesh Singh. 2020. "A Unified Framework for Big Data Acquisition, Storage, and Analytics for Demand Response Management in Smart Cities." *Future Generation Computer Systems* 108: 921–934.

Jirn, Jin Suh. 2015. "A Sort of European Hallucination." *Situations* 8 (2): 67–83.

Kaku, Michio. 2023. *Quantum Supremacy: How the Quantum Computer Revolution Will Change Everything*. New York: Doubleday.

Kalpokas, Ignas. 2022. "Posthuman Urbanism: Datafication, Algorithmic Governance and Covid-19." In *The Routledge Handbook of Architecture, Urban Space and Politics*, vol. I, 496–508. London: Routledge, Taylor and Francis Group.

Kaplan, Robert. 1966. "Cultural Thought Patterns in Inter-Cultural Education." *Language Learning* 16 (1–2): 1–20.

Kauffman, Stuart, and Andrea Roli. 2022. "What Is Consciousness? Artificial Intelligence, Real Intelligence, Quantum Mind and Qualia." *Biological Journal of the Linnean Society* 20: 1–9.

Keller, Evelyn Fox. 1985. *Reflections on Gender and Science*. New Haven, CT: Yale University Press.

Kerenyi, Karl. 1951. *The Gods of the Greeks*. London: Thames and Hudson.

Kitchin Rob. 2014. "Big Data, New Epistemologies and Paradigm Shifts." *Big Data & Society* 1 (1). https://doi.org/10.1177/2053951714528481.

Koek, Ariane. 2020. "Matter, Space and Time: The Enchanting Entanglement of Physics and Art." *Quarks Daily*. https://3quarksdaily.com/3quarksdaily/2020/02/matter-space-and-time-the-enchanting-entanglement-of-physics-and-art.html.

Koerner, Konrad. 1992. "The Sapir-Whorf Hypothesis: A Preliminary History and a Bibliographical Essay." *Journal of Linguistic Anthropology* 2 (2): 173–198.

Korte, Martin. 2022. "The Impact of the Digital Revolution on Human Brain and Behavior: Where Do We Stand?" *Dialogues in Clinical Neuroscience* 22 (2): 101–111.

Kotka, Taavi, Carlos Ivan Vargas Alvarez del Castillo, and Kaspar Korjus. 2015. "Estonian E-Residency: Redefining the Nation-State in the Digital Era." *Cyber Studies Programme Working Paper*, Series 3. https://www.ctga.ox.ac.uk/article/estonian-e-residency-redefining-nation-state-digital-era.

Kuhn, Thomas. 1970. *The Structure of Scientific Revolutions*. 2nd ed. Chicago: University of Chicago Press.

Kumar, Manjit. 2008. *Quantum: Einstein, Bohr and the Great Debate about the Nature of Reality*. London: Icon Books Ltd.

Kummitha, Rama Krishna Reddy. 2020. "Why Distance Matters: The Relatedness between Technology Development and Its Appropriation in Smart Cities." *Technological Forecasting and Social Change* 157: 120087.

Labatout, Benjamin. 2021. *When We Ceased to Understand the World*. London: Pushkin.

Lai, Vicky, and Lera Boroditsky. 2013. "The Immediate and Chronic Influence of Spatio-Temporal Metaphors on the Mental Representations of Time in English, Mandarin, and Mandarin-English Speakers." *Frontiers in Psychology* 4: 142.

Lake, Robert. 1993. "Planning and Applied Geography: Positivism, Ethics, and Geographic Information Systems." *Progress in Human Geography* 17: 404–413.

Lakoff, George, and Mark Johnson. 1999. *Philosophy in the Flesh: The Embodied Mind and its Challenge to Western Thought*. New York: Basic Books.

Landau, Martin. 1969. "Redundancy, Rationality, and the Problem of Duplication and Overlap." *Public Administration Review* 29 (4): 346–358.

Landauer, Rolf. 1996. "The Physical Nature of Information." *Physics Letters A* 217 (4–5): 188–193.

Latour, Bruno. 1987. *Science in Action: How to Follow Scientists and Engineers through Society*. Cambridge, MA: Harvard University Press.

Lazer, David, Ryan Kennedy, Gary King, and Alessandro Vespignani. 2014. "The Parable of Google Flu: Traps in Big Data Analysis." *Science* 343 (6176): 1203–1205.

Lee, Kai-Fu. 2018. *AI Superpowers: China, Silicon Valley, and the New World Order*. New York: Houghton Mifflin Harcourt.

Lemmens, Pieter. 2011. "An Interview with Bernard Stiegler." *Krisis* 1: 33–41.

Leroi-Gourhan, André. 1993. *Gesture and Speech*. Translated by Anna Bostock Berger. Cambridge, MA: MIT Press.

Libet, Benjamin. 1999. "Do We Have Free Will?" *Journal of Consciousness Studies* 6 (8–9): 47–57.

Logan, Robert. 2004. *The Alphabet Effect: A Media Ecology Understanding of the Making of Western Civilization*. New York: Hampton Press.

LSE Truth, Trust & Technology Commission. 2019. "Tackling the Information Crisis: A Policy Framework for Media System Resilience." *London School of Economics and Political Science*. https://www.lse.ac.uk/media-and-communications/assets /documents/research/T3-Report-Tackling-the-Information-Crisis.pdf

Lupyan, Gary, and Rick Dale. 2016. "Why Are There Different Languages? The Role of Adaptation in Linguistic Diversity." *Trends in Cognitive Sciences* 20 (9): 649–660.

Lyon, David. 2007. *Surveillance Studies. An Overview*. Cambridge: Polity Press.

Lyotard, François. 1984. *The Postmodern Condition: A Report on Knowledge*. Translated by Geoff Bennington and Brian Massumi. Minneapolis: Minnesota University Press.

MacNeilage, Peter. 2008. *The Origin of Speech*. Oxford: Oxford University Press.

MacSweeney, Mairéad, Cheryl M. Capek, Ruth Campbell, and Bencie Woll. 2008. "The Signing Brain: The Neurobiology of Sign Language." *Trends in Cognitive Sciences* 12 (11): 432–440.

Marlow, A. R., ed. 1978. *Mathematical Foundations of Quantum Theory*. New Orleans: Academic Press.

Masuda, Takahiko, and Richard Nisbett. 2001. "Attending Holistically vs. Analytically: Comparing the Context Sensitivity of Japanese and Americans." *Journal of Personality and Social Psychology* 81: 922–934.

Massumi, Brian. 2017. "Virtual Ecology and the Question of Value." In *General Ecology: The New Ecological Paradigm*, edited by Erich Hörl with James Burton, 345–374. London: Bloomsbury.

Mattern, Shannon. 2021. *A City Is Not a Computer: Other Urban Intelligences*. Princeton, NJ: Princeton University Press.

Maturana, Humberto, and Francisco Varela. 1980. *Autopoiesis and Cognition: The Realization of the Living*. 2nd ed. Dordrecht: Springer.

Maturana, Humberto, and Francisco Varela. 1987. *The Tree of Knowledge: The Biological Roots of Human Understanding*. Boston: Shambhala Publications.

McGilchrist, Iain. 2009. *The Master and His Emissary. The Divided Brain and the Making of the Western World*. New Haven, CT: Yale University Press.

McLuhan, Eric, and Frank Zingrone. 1995. *Essential McLuhan*. New York: Basic Books.

McLuhan, Marshall. 1962. *The Gutenberg Galaxy: The Making of Typographic Man*. Toronto: Toronto University Press.

McLuhan, Marshall. 1964. *Understanding Media*. New York: McGraw Hill.

McLuhan, Marshall. 1989. *The Man and His Message*. Edited by George Sanderson and Frank Macdonald. Golden, CO: Fulcrum Publishing.

McLuhan, Marshall, and Robert Logan. 1977. "Alphabet, Mother of Invention." *ETC: A Review of General Semantics* 34: 373–383.

Meighoo, Sean. 2008. "Derrida's Chinese Prejudice." *Cultural Critique* 68: 163–209.

Mejias, Ulises. 2013. *Off the Network: Disrupting the Digital World*. Minneapolis: University of Minnesota Press.

Mengyu, Li. 2008. "The Unique Values of Chinese Traditional Cultural Time Orientation: In Comparison with Western Cultural Time Orientation." *Intercultural Communication Studies* 1: 64–70.

Merletto, Chiara. 2022. *La scienza dell'impossibile. Alla ricerca delle nuove leggi della fisica*. Milano: Mondadori.

Metcalf, Jacob, and Kate Crawford. 2016. "Where Are Human Subjects in Big Data Research? The Emerging Ethics Divide." *Big Data & Society* 3 (1). https://doi.org/10.1177/2053951716650211.

Meyrowitz, Joshua. 1999. "No Sense of Place: The Impact of Electronic Media on Social Behaviour." In *The Media Reader: Continuity and Transformation*, edited by Hugh Mackay and Tim O'Sullivan, 99–120. London: Sage.

Mitchell, William John. 1994. *Picture Theory: Essays on Verbal and Visual Representation*. Chicago: University of Chicago Press.

Miyamoto, Yuri, Richard Nisbett, and Takahiko Masuda. 2006. "Culture and the Physical Environment. Holistic versus Analytic Perceptual Affordances." *Psychological Science* 17 (2): 113–119.

Mudrooroo, Narogin. 1995. *Us Mob: History, Culture, Struggle: An Introduction to Indigenous Australia*. Sydney: Angus & Robertson.

Mueller, Milton. 2019. "Sovereignty and Cyberspace: Institutions and Internet Governance." https://dlc.dlib.indiana.edu/dlc/bitstream/handle/10535/10410/5th-Ostrom-lecture-DLC.pdf?sequence=1&isAllowed=y.

Mukai, Hiroto, et al. 2020. "Pseudo-2D Superconducting Quantum Computing Circuit for the Surface Code: Proposal and Preliminary Tests." *New Journal of Physics* 22: 043013. https://iopscience.iop.org/article/10.1088/1367-2630/ab7d7d.

Mutsvairo, Bruce, and Massimo Ragnedda, eds. 2019. *Mapping the Digital Divide in Africa: A Mediated Analysis*. Amsterdam: Amsterdam University Press.

Needham, Joseph. 1954. *Science and Civilisation in China*. Cambridge: Cambridge University Press.

Newen, Albert, Leon De Bruin, and Shaun Gallagher, eds. 2018. *The Oxford Handbook of 4E Cognition*. Oxford: Oxford University Press.

Newton, Isaac. 1687. *Philosophiae Naturalis Principia Mathematica*. Londini, Jussu Societatis Regiæ ac Typis Josephi Streater. Prostat apud plures Bibliopolas.

Ng, Melvin, et al. 2017. "How We Think about Temporal Words: A Gestural Priming Study in English and Chinese." *Frontiers in Psychology* 8: 974.

Nicolescu, Basarab. 2005. "Towards Transdisciplinary Education." *The Journal for Transdisciplinary Research in Southern Africa* 1 (1): 5–16.

Nisbett, Richard. 2003. *The Geography of Thought: How Asians and Westerners Think Differently—And Why*. New York: Free Press.

Nochta, Timea, Li Wan, Jennifer Mary Schooling, and Ajith Kumar Parlikad 2021. "A Socio-technical Perspective on Urban Analytics: The Case of City-scale Digital Twins." *Journal of Urban Technology* 28 (1–2): 263–287.

Nonnecke, Brandie. 2016. "The Transformative Effects of Multistakeholderism in Internet Governance: A Case Study of the East Africa Internet Governance Forum." *Telecommunications Policy* 40 (4): 343–352.

Ong, Walter. 1986. "Writing Is a Technology That Restructures Thought." In *The Written Word: Literacy in Transition*, edited by Gerd Baumann, 23–50. New York: Clarendon Press.

Ostrom, Elinor. 2010. "Beyond Markets and States: Polycentric Governance of Complex Economic Systems." *American Economic Review* 100 (3): 641–672.

Pae, Hye K. 2020. *Script Effects as the Hidden Drive of the Mind, Cognition, and Culture*. Cham: Spring Nature.

Pashler, Harold. 1994. "Dual-task Interference in Simple Tasks: Data and Theory." *Psychological Bulletin* 116 (2): 220.

Parikka, Jussi. 2014. *The Anthrobscene*. Minneapolis: University of Minnesota Press.

Pauka, Sebastian, et al. 2021. "A Cryogenic CMOS Chip for Generating Control Signals for Multiple Qubits." *Nature Electronics* 4: 64–70.

Penrose, Roger. 1994. *Shadows of the Mind: A Search for the Missing Science of Consciousness*. Oxford: Oxford University Press.

Perez, Carlota. 2010. "Technological Revolutions and Techno-economic Paradigms." *Cambridge Journal of Economics* 34 (1): 185–202.

Perrier, Elija. 2021. "Ethical Quantum Computing: A Roadmap." *arXiv preprint*. https://arxiv.org/pdf/2102.00759.pdf.

Perrier, Elija. 2022. "The Quantum Governance Stack: Models of Governance for Quantum Information Technologies." *Digital Society* 1 (3): 22.

Petersen, Aage. 1985. "The Philosophy of Niels Bohr." In *Niels Bohr: A Centenary Volume*, edited by Anthony French and P. J. Kennedy. Cambridge, MA: Harvard University Press.

Piaget, Jean. 1936. *Origins of Intelligence in the Child*. London: Routledge.

Pierrehumbert, Janet, Mary Esther Beckman, and Robert Ladd. 2000. "Conceptual Foundations of Phonology as a Laboratory Science." In *Phonological Knowledge: Its Nature and Status*, edited by Noel Burton-Roberts, Philip Carr, and Gerard Docherty, 273–303. Cambridge: Cambridge University Press.

Pinker, Steven. 1994. *The Language Instinct*. New York: W. Morrow and Co.

Plotnitsky, Arkady. 1994. *Complementarity: Anti-epistemology after Bohr and Derrida*. Durham, NC: Duke University Press.

Polanyi, Michael. (1966) 2009. *The Tacit Dimension*. Chicago: University of Chicago Press.

Portugali, Juval. 2011. *Complexity, Cognition and the City*. Dordrecht: Springer Science & Business Media.

Possati, Luca M. 2023. "Ethics of Quantum Computing: An Outline." *Philosophy & Technology* 36 (3): 48. https://link.springer.com/article/10.1007/s13347-023-00651-6.

Postman, Neil. 1970. "The Reformed English Curriculum." In *High School 1980: The Shape of the Future in American Secondary Education*, edited by Alvin C. Eurich, 160–168. New York: Pitman.

Pouw, Wim, and Susanne Fuchs. 2022. "Origins of Vocal-Entangled Gesture." *Neuroscience & Biobehavioral Reviews* 141: 104836.

Preskill, John. 2012. "Quantum Computing and the Entanglement Frontier." *arXiv preprint*. arXiv:1203.5813.

Priest, Graham. 2007. "Paraconsistency and Dialetheism." In *Handbook of the History of Logic*, vol. 8, edited by Dov Gabby and John Woods, 129–204. Amsterdam: Elsevier.

Priest, Graham, Jeffrey C. Beall, and Bradley Armour-Garb, eds. 2006. *The Law of Non-Contradiction: New Philosophical Essays*. Oxford: Oxford University Press.

Proietti, Massimiliano, et al. 2019. "Experimental Test of Local Observer Independence." *Science Advances* 5 (9). https://doi.org/10.1126/sciadv.aaw9832.

Purtova, Nadezhda. 2017. "Health Data for Common Good: Defining the Boundaries and Social Dilemmas of Data Commons." In *Under Observation: The Interplay between eHealth and Surveillance*, edited by Samantha Adams, Nadezhda Purtova, and Ronald Leenes, 177–210. Berlin: Springer.

Purvis, Ben, Yong Mao, and Darren Robinson. 2019. "Three Pillars of Sustainability: In Search of Conceptual Origins." *Sustainability Science* 14, 681–695.

Pusey, Matthew F., Jonathan Barrett, and Terry Rudolph. 2012. "On the Reality of the Quantum State." *Nature Physics* 8 (6): 475–478.

Quartiroli, Ivo. 2011. *The Digitally Divided Self: Relinquishing Our Awareness to the Internet*. Milano: Silens.

Quijano, Aníbal. 2007. "Coloniality and Modernity/Rationality." *Cultural Studies* 21 (2–3): 168–178.

Randall, Bob. 2003. *Songman: The Story of an Aboriginal Elder of Uluru*. Sydney: ABC Books for the Australian Broadcasting Corporation.

Random, Michel. 1985. *La stratégie de l'invisible*. Paris: Félin.

Rantanen, Terhi. 2006. "A Man behind Scapes: An Interview with Arjun Appadurai." *Global Media and Communication* 2 (1): 7–19.

Reeh Peters, Christine. 2023. "Film as Artificial Intelligence: Jean Epstein, Film-Thinking and the Speculative-Materialist Turn in Contemporary Philosophy." *Film-Philosophy* 27 (2): 151–172. https://doi.org/10.3366/film.2023.0224.

Rizzo, Susanna, and Greg Melleuish. 2021. "In Search of the Origins of the Western Mind: McGilchrist and the Axial Age." *Histories* 1: 24–41.

Roberts, Huw, et al. 2020. "The Chinese Approach to Artificial Intelligence: An Analysis of Policy, Ethics, and Regulation." *AI & Society* 36: 59–77.

Roques-Carmes, Charles, et al. 2023. "Biasing the Quantum Vacuum to Control Macroscopic Probability Distributions." *Science* 381 (6654): 205–209.

Rosenberg, Louis B. 2022. "Regulating the Metaverse, a Blueprint for the Future." In *Proceedings of the International Conference on Extended Reality*, edited by Lucio Tommaso De Paolis, Pasquale Arpaia, and Marco Sacco, 263–272. Cham: Springer.

Rošker, Jana. 2021. "Epistemology in Chinese Philosophy." In *The Stanford Encyclopaedia of Philosophy*, edited by Edward Zalta. https://plato.stanford.edu/entries/chinese-epistemology/.

Rovelli, Carlo. 1996. "Relational Quantum Mechanics." *International Journal of Theoretical Physics* 35 (8): 1637–1678.

Rovelli, Carlo. 2020. *Helgoland*. Milano: Adelphi.

Sampson, Geoffrey. 2015. *Writing Systems*. Bristol: Equinox.

Saracco, Roberto. 2018. "Transhumanism." *IEEE—Future Direction Committee Symbiotic Autonomous Systems Initiative*. https://digitalreality.ieee.org/images/files/pdf/transhumanism.pdf.

Sarraute, Nathalie. 1939. *Tropismes*. Paris: Editions Denoël.

Sarraute, Nathalie. 1956. *L'ère du soupçon*. Paris: Nouvelle Revue Française.

Schmandt-Besserat, Denise. 1977. "An Archaic Recording System and the Origin of Writing." *Syro-Mesopotamian Studies* 1 (2): 1–32.

Schmid, Hans Bernard. 2014. "Plural Self-Awareness." *Phenomenology and the Cognitive Sciences* 13 (1): 7–24.

Schneider, Ingrid. 2020. "Democratic Governance of Digital Platforms and Artificial Intelligence? Exploring Governance Models of China, the US, the EU and Mexico." *JeDEM—eJournal of eDemocracy and Open Government* 12 (1): 1–24.

Schrödinger, Erwin. 1957. *Science, Theory, and Man*. New York: Dover.

Sennett, Richard. 2018. *Building and Dwelling: Ethics for the City*. New York: Farrar, Straus and Giroux.

Serfati, Michel. 2005. *La révolution symbolique, la constitution de l'écriture symbolique mathématique*. Paris: Pétra.

Shapin, Steven. 1994. *A Social History of Truth: Civility and Science in Seventeenth-Century England*. Chicago: University of Chicago Press.

Shahat, Ehab, Chang Hyun, and Chunho Yeom. 2021. "City Digital Twin Potentials: A Review and Research Agenda." *Sustainability* 13 (6): 3386.

Shaviro, Steve. 1997. *Doom Patrols: A Theoretical Fiction about Postmodernism*. New York: Serpent's Tail.

Shimony, Abner. 1997. *On Mentality, Quantum Mechanics, and the Actualization of Potentialities*. Cambridge: Cambridge University Press.

Shrivastava, Paul, Mark Stafford Smith, Karen O'Brien, and Laszlo Zsolnai. 2020. "Transforming Sustainability Science to Generate Positive Social and Environmental Change Globally." *One Earth* 2 (4): 329–340.

Silver, Gerald, and Myma Silver. 1989. *Systems Analysis and Design*. Boston: Addison Wensley.

Simanowski, Roberto. 2011. *Digital Art and Meaning. Reading Kinetic Poetry, Text Machines, Mapping Art and Interactive Installations*. Minneapolis: University of Minnesota Press.

Simanowski, Roberto. 2019. "On the Ethics of Algorithmic Intelligence." *Social Research* 86 (2): 423–447.

Sloterdijk, Peter. 2000. *La domestication de l'être*. Paris: Mille et une nuits.

Smith, Frank. 2020. "Quantum Technology Hype and National Security." *Security Dialogue* 51 (5): 499–516.

Smolin, Lee. 2019. *Einstein's Unfinished Revolution: The Search for What Lies beyond the Quantum*. New York: Penguin.

Soini, Katriina, and Joost Dessein. 2016. "Culture-Sustainability Relation: Towards a Conceptual Framework." *Sustainability* 8 (167): 1–12.

Sombart, Werner. 1987. *Der moderne Kapitalismus: Historischsystematische Darstellung des gesamteuropäischen Wirtschaftslebens von seinen Anfängen bis zur Gegenwart, vol. 2, pt. 1: Das europäische Wirtschaftsleben im Zeitalter der Frühkapitalismus, vornehmlich im 16., 17. und 18. Jahrhundert*. München: Deutscher Taschenbuch Verlag.

Sörlin, Sverker, and Nina Wormbs. 2018. "Environing Technologies: A Theory of Making an Environment." *History and Technology* 34: 101–125.

Spargo, John. 1913. *Syndicalism, Industrial Unionism and Socialism*. New York: Huebsch.

Sperber, Dan. 1990. "The Epidemiology of Beliefs." In *The Social Psychological Study of Widespread Beliefs*, edited by Colin Fraser and George Gaskell, 25–44. Oxford: Clarendon.

Stiegler, Bernard. 1993. *Technics and Time: The Fault of Epimetheus*. Translated by Richard Beardsworth. Palo Alto, CA: Stanford University Press.

Stiegler, Bernard. 2009. *Technics and Time: Disorientation*. Translated by Stephen Barker. Palo Alto, CA: Stanford University Press.

Stiegler, Bernard. 2015. *States of Shock: Stupidity and Knowledge in the Twenty-first Century*. London: Polity.

Stiegler, Bernard. 2016. "The Digital, Education, and Cosmopolitanism." *Representation* 134: 157–164.

Stiegler, Bernard. 2017. "General Ecology, Economy, and Organology." In *General Ecology: The New Ecological Paradigm*, edited by Erich Hörl with James Burton, 129–150. London: Bloomsbury.

Stout, Dietrich, and Thierry Chaminade. 2012. "Stone Tools, Language and the Brain in Human Evolution." *Philosophical Transactions of the Royal Society B* 367: 75–87.

Strassberg, Richard. 1994. *Inscribed Landscapes: Travel Writing from Imperial China*. Berkeley: University of California Press.

Sun, Pei. 2007. "Is the State-Led Industrial Restructuring Effective in Transition China? Evidence from the Steel Sector." *Cambridge Journal of Economics* 31 (4): 601–624.

Susskind, Jamie. 2022. *The Digital Republic: On Freedom and Democracy in the 21st Century*. Sydney: Bloomsbury Publishing.

Tacchino, Francesco, Chiara Macchiavello, Dario Gerace, and Daniele Bajoni. 2019. "An Artificial Neuron Implemented on an Actual Quantum Processor." *npj Quantum Information* 5. https://doi.org/10.1038/s41534-019-0140-4

Taylor, Charles. 1986. *Philosophical Papers I: Human Agency and Language*. Cambridge: Cambridge University Press.

Taylor, Insup, and Maurice Martin Taylor. 2014. *Writing and Literacy in Chinese, Korean, and Japanese*. Amsterdam: John Benjamins.

Taylor, Linnet. 2017. "What Is Data Justice? The Case for Connecting Digital Rights and Freedoms Globally." *Big Data & Society* 4 (2). https://doi.org/10.1177/2053951717736335.

Taylor, Linnet. 2020. "The Price of Certainty: How the Politics of Pandemic Data Demand an Ethics of Care." *Big Data & Society*, July–December, 1–7. https://doi.org/10.1177/2053951720942539.

Taylor, Linnet, and Dennis Broeders. 2015. "In the Name of Development: Power, Profit and the Datafication of the Global South." *Geoforum* 64 (August): 229–237.

Taylor, Richard D. 2020. "Quantum Artificial Intelligence: A 'Precautionary' U.S. Approach?" *Telecommunications Policy* 4 (6): 101909. https://doi.org/10.1016/j.telpol.2020.101909.

Theise, Neil, and Menas Kafatos. 2013. "Complementarity in Biological Systems: A Complexity View." *Complexity* 16 (6): 11–20.

Thompson, Clive. 2013. *Smarter Than You Think: How Technology Is Changing Our Minds for the Better*. London: Penguin.

Tomasello, Michael. 2008. *Origins of Human Communication*. Cambridge, MA: MIT Press.

Toots, Maarja. 2019. "Why e-Participation Systems Fail: The Case of Estonia's Osale.ee." *Government Information Quarterly* 36 (3): 546–559.

Vafopoulos, Michalis. 2012. "Being, Space, and Time on the Web." *Metaphilosophy* 43 (4): 405–425.

van Lente, Harro. 2000. "Forceful Futures: From Promise to Requirement." In *Contested Futures: A Sociology of Prospective Techno-Science*, edited by Nik Brown and Brian Rappert, 43–64. New York: Ashgate.

van Nimwegen, Christof. 2008. *The Paradox of the Guided User: Assistance Can Be Counter-effective*. Utrecht: Universiteit Utrecht.

Vannoy Adams, Michael. 2012. "The Archetype of the Saboteur: Self-Sabotage from a Jungian Perspective." http://www.jungnewyork.com/archetype-saboteur.shtml.

Varela, Francisco, Evan Thompson, and Eleanor Rosch. 1991. *The Embodied Mind: Cognitive Science and Human Experience*. Cambridge, MA: MIT Press.

Velmans, Max. 2000. *Understanding Consciousness*. London: Routledge.

Vermaas, Pieter. 2017. "The Societal Impact of the Emerging Quantum Technologies: A Renewed Urgency to Make Quantum Theory Understandable." *Ethics and Information Technology* 19: 241–246.

Vico, Giambattista. 1744. *The New Science of Giambattista Vico.* Translated by Thomas G. Bergin and Max H. Fisch. Ithaca, NY: Cornell University Press.

Vigneau, Mathieu, et al. 2011. "What Is Right-Hemisphere Contribution to Phonological, Lexico-Semantic, and Sentence Processing? Insights from a Meta-Analysis." *NeuroImage* 54: 577–593.

Viitanen, Jenni, and Richard Kingston. 2014. "Smart Cities and Green Growth: Outsourcing Democratic and Environmental Resilience to the Global Technology Sector." *Environment and Planning A* 46: 803–819.

von der Leyen, Ursula. 2020. "A Union that Strives for More: My Agenda for Europe." https://ec.europa.eu/info/sites/default/files/political-guidelines-next-commission_en_0.pdf.

von Neumann, John. (1955) 1986. "John von Neumann on Technological Prospects and Global Limits." *Population and Development Review* 12 (1): 117–126.

von Uexküll, Jacob. (1934) 2010. *A Foray into the Worlds of Animals and Humans with a Theory of Meaning.* Translated by Joseph D. O'Neill. Minneapolis: University of Minnesota Press.

Voßkuhle, Andreas. 2008. "Das Konzept des rationalen Staates." In *Governance von und durch Wissen,* edited by Gunnar Folke Schuppert and Andreas Voßkuhle, 11–33. Baden-Baden: Nomos.

Vygotsky, Lev. 1986. *Thought and Language.* Cambridge, MA: MIT Press.

Walsh, Denis M. 2015. *Organisms, Agency, and Evolution.* Cambridge: Cambridge University Press.

Wambacq, Judith, and Bart Buseyne. 2012. "The Reality of Real Time." *New Formations* 77: 63–75.

Wang, Min, Mingke Rao, and Zhipeng Sun. 2020. "Typology, Etiology, and Fact-Checking: A Pathological Study of Top Fake News in China." *Journalism Practice* 16 (4): 719–737.

Wang, Yingxu. 2013. "Cognitive Linguistic Perspectives on the Chinese Language." *New Mathematics and Natural Computation* 9 (2): 237–260.

Weitz, Nina, Hendrik Carlsen, Mans Nilsson, and Kristian Skanberg. 2018. "Towards Systemic and Contextual Priority Setting for Implementing the 2030 Agenda." *Sustainable Science* 13: 531–548.

Wendt, Alexander. 2015. *Quantum Mind and Social Science.* Cambridge: Cambridge University Press.

Wheeler, John. 1979. "Frontiers of Time." In *Problems in the Foundations of Physics, Proceedings of the International School of Physics "Enrico Fermi,"* edited by G. Toraldo di Francia. New York: North-Holland.

White, Sara, Rebecca Johnson, Simon Liversedge, and Keith Rayner. 2008. "Eye Movements When Reading Transposed Text: The Importance of Word-Beginning Letters." *Journal of Experimental Psychology* 34: 1261–1276.

Wigner, Eugene, P. 1961. "Remarks on the Mind-Body Question." In *Quantum Theory and Measurement*, edited by John A. Wheeler and Wojciech H. Zurek, 168–181. Princeton, NJ: Princeton University Press.

Winseck, Dwayne. 2017. "The Geopolitical Economy of the Global Internet Infrastructure." *Journal of Information Policy* 7: 228–267.

Wittgenstein, Ludwig. 2016. *Interactive Dynamic Presentation (IDP) of Ludwig Wittgenstein's Philosophical Nachlass*, edited by the Wittgenstein Archives at the University of Bergen under the direction of Alois Pichler. https://wab.uib.no/transform/transformer.php.

Wolf, Maryanne. 2007. *Proust and the Squid: The Story and Science of the Reading Brain.* New York: Harper.

WRR. 2021. *Opgave AI. De nieuwe systeemtechnologie*. The Hague: WRR.

Yates, Francis. 1966. *The Art of Memory*. London: Routledge.

Young, Thomas. 1807. *A Course of Lectures on Natural Philosophy and the Mechanical Arts*. Vol. 1. London: Johnson.

Yushu, Gong. 2010. "The Sumerian Account of the Invention of Writing: A New Interpretation." *Procedia Social and Behavioral Sciences* 2: 7446–7453.

Zabaleta, Omar Gustavo, Juan Pablo Barrangú, and Constancio M. Arizmendi. 2017. "Quantum Game Application to Spectrum Scarcity Problems." *Physica A: Statistical Mechanics and Its Applications* 466: 455–461.

Zeilinger, Anton. 1996. "On the Interpretation and Philosophical Foundation of Quantum Mechanics." In *Vastakohtien Todellisuus: Festschrift for K. V. Laurikainen*, edited by Urho Ketvel et al. Helsinki: Helsinki University Press.

Zellini, Paolo. 2022. *Discreto e continuo. Storia di un errore*. Milano: Adelphi.

Zhou, Rong. 2001. "The Psychological Reality of Cognitive Basis of Metaphors: Evidence from the Spatial Metaphoric Representation of Time." *Foreign Language Teaching Research* 33: 88–93.

Zohar Danah, and Ian Marshall. 1994. *The Quantum Society: Mind, Physics, and a New Social Vision*. New York: Bloomsbury.

Zuboff, Shoshana. 2013. "The Surveillance Paradigm: Be the Friction—Our Response to the New Lords of the Ring." *Frankfurter Allgemeine Zeitung*, June 26. http://www.faz.net/aktuell/feuilleton/the-surveillance-paradigm-be-the-friction-our-response-to-the-new-lords-of-the-ring-12241996.html.

Zuboff, Shoshana. 2019. *The Age of Surveillance Capitalism*. London: Profile Books.

Index